Uniscience Series on Fine Particle Science and Technology

Editor

John Keith Beddow, Ph.D.
Professor of Chemicals and Materials Engineering
Division of Materials Engineering
University of Iowa
Iowa City, Iowa

Advanced Particulate Morphology
J. K. Beddow and T. P. Meloy

Separation of Particles From Air and Gases, Volumes I and II
Akira Ogawa

Particle Characterization in Technology
Volume I: Applications and Microanalysis
Volume II: Morphological Analysis
J. K. Beddow, Editor

Separation of Particles From Air and Gases

Volume I

Author

Akira Ogawa, Dr. Eng.
Associate Professor
Department of Mechanical Engineering
College of Engineering
Nihon University
Fukushima, Japan

Uniscience Series on Fine Particle Science and Technology

Editor-in-Chief

John Keith Beddow, Ph.D.
Professor of Chemicals and Materials Engineering
Department of Materials Engineering
University of Iowa
Iowa City, Iowa

CRC Press, Inc.
Boca Raton, Florida

Library of Congress Cataloging in Publication Data

Ogawa, Akira, 1940 —
    Separation of particles from air and gases.

    (Uniscience series on fine particle science technology)
    Includes bibliographical references and index.
    1. Dust—Removal—Equipment and supplies. 2. Dust
control. 3. Air—Cleaning. 4. Gases—Cleaning.
I. Title. II. Series.
TH7692.035 1984   628.5'3   82-20679
ISBN 0-8493-5787-X (v.1)
ISBN 0-8493-5788-8 (v.2)

    Direct all inquiries to CRC Press, Inc., 2000 Corporate Blvd., N.W., Boca Raton, Florida, 33431.

© 1984 by CRC Press, Inc.

International Standard Book Number 0-8493-5787-X (Volume 1)
International Standard Book Number 0-8493-5788-8 (Volume 2)

Library of Congress Card Number 82-20679

Printed in the United States

PREFACE

Dr. John Aitken wrote in his paper (On the formation of small clear spaces in dusty air, *Trans. R. Soc. Eng.*, 32, 1883-1884) the following:

The dust particles floating in our atmosphere are every day demanding more and more attention. As our knowledge of these unseen particles increases, our interest deepens, and I might almost say gives place to anxiety, when we realize the vast importance these dust particles have on life, whether it be those inorganic ones so small as to be beyond the powers of the microscope, or those larger organic ones which float unseen through our atmosphere, and which, though invisible, are yet the messengers of sickness and of death to many — messengers far more real and certain than poet or painter has ever conceived . . .

Recently, with developments in science and technology, physical and chemical properties of solid particles, dust, and fumes have become clearer in comparison to the knowledge of particle properties of about 50 or 100 years ago. Dust collectors for air pollution and the chemical and mechanical industries must be planned corresponding to the new knowledge of these dust properties.

In order to design new types of dust collectors, it is necessary not only to rely on usual experience and techniques, but also to grasp the fundamental mechanisms and behaviors of the fine solid particles in the turbulent gas flow occurring in ducts or separation chambers. Then, in order to urge the new creative ideas for the design of dust collectors, the author describes the motion of coarse and fine solid particles in turbulent gas flows in detail. Further, from the fluid dynamical point of view, many kinds of the fundamental constructions of dust collectors are shown with many illustrations.

THE EDITOR-IN-CHIEF

John Keith Beddow received his Ph.D. in Metallurgy from Cambridge University, Cambridge, England, in 1959. Currently Secretary of the Fine Particle Society, he is a member of the Faculty at the University of Iowa, where he heads a small research group in fine particle science with emphasis on morphological analysis. Dr. Beddow is an active lecturer and author. He has also been active as a Consultant in metallurgy, powder metallurgy, and powder technology for numerous corporations. He is also president of Shape Technology, Ltd. His present research activities are in particle morphological analysis. Dr. Beddow is married, with four daughters, and has resided in the U.S. since 1966.

THE AUTHOR

Akira Ogawa, Dr. Eng., was born in 1940 in Tokyo and is currently Associate Professor, Department of Mechanical Engineering, College of Engineering, Nihon University, Fukushima, Japan.

He received a Bachelor's degree in Mechanical Engineering from the College of Science and Technology, Nihon University in 1963, a Masters in Engineering from Nihon University in 1965, and a Doctorate in Engineering from Nihon University in 1972.

He worked as an assistant in and as an Assistant Professor at the Department of Mechanical Engineering in the College of Science and Technology, Nihon University from 1968 to 1973. He moved to the Department of Mechanical Engineering, College of Engineering, Nihon University as an Associate Professor in 1974.

Dr. Ogawa is a member of the Japan Society of Mechanical Engineering, the Japan Society of Chemical Engineering, the Japan Society of Powder Technology, the Japan Society of Precision Engineering, the Japan Society of Air Pollution, the Japan Society of Fluid Mechanics, and the Fine Particle Society (U.S.A.).

Dr. Ogawa is the author of *Cyclone Dust Collectors* and *Vortex Flow* (both in Japanese).

His major research interests include the turbulent rotational flow in the confined vortex chamber, the Taylor-Görtler vortex flow in the coaxial cylindrical chamber, turbulent swirling air jets, cyclone dust collectors, rotary flow dust collectors, and the behavior of fine solid particles in turbulent air flow.

ACKNOWLEDGMENTS

The author wishes to express his gratitude to Prof. Dr. John Keith Beddow of the University of Iowa for his advice, encouragement, and suggestions.

Further, it is a pleasure to acknowledge the contributions of Prof. Dr. I. Tani of Nihon University, Tokyo; Prof. Dr. K. Iinoya of Aichi K. University, Nagoya; Dipl.-Ing. H. Klein and Dipl.-Phys. R. Pieper of Siemens-Schuckertwerken, Erlangen; Prof. Dr. K. Melcher, Dipl.-Ing. J. Komaroff, and Dipl.-Ing. K. Ito of Technisches Zentrum; Robert Bosh, Stuttgart; Dr. M. P. Escudier of Brown Boveri, Baden; Prof. Dr. E. Muschelknautz of Bayer AG, Dormagen; Prof. Dr. H. Brauer of Technische Universität, Berlin; Prof. Dr. M. Bohnet of Technische Universität, Braunschweigh; Prof. Dr. F. Löffler of Technische Universität, Karlsruhe; Dr. J. C. Rotta of DFVLR, Göttingen; Prof. Dr. O. Molerus of Technische Universität, Erlangen-Nürnberg; Prof. Dr. C. Alt of Technische Universität, Stuttgart; Prof. Dr. J. O. Hinze of Delft University, Delft; Prof. Dr. A. Fortier, Prof. Dr. R. Comolet, and Dr. C. P. Chen of the University of Paris (VI); Prof. Dr. J. Abrahamson of Canterbury University, New Zealand; Dr. W. H. Gauvin of Noranda, Québec; Prof. Dr. P. L. Corbella of the University of Barcelona, and Prof. Dr. E. Costa Novella of the University of Madrid, Spain; and Prof. Dr. R. H. Page, of Texas A & M University.

The author should like to express his gratitude to the following authors:

Beddow, J. K., *Particulate Science and Technology*, Chemical Publishing, New York, 1982.

Bethea, R. M., *Air Pollution Control Technology*, D. Van Nostrand, New York, 1978.

Blacktin, S. C., *Dust*, Chapman & Hall, London, 1934.

Brauer, H. and Varma, Y. B. G., *Air Pollution Control Equipment*, Springer-Verlag, Berlin, 1981.

Comolet, R., Dynamiques des fluides réeles, *Tome II*, Masson et Cie, Paris, 1982.

Crawford, M., *Air Pollution Control Theory*, McGraw-Hill, New York, 1976.

Davies, C. N., *Aerosol Science*, Academic Press, New York, 1966.

Dorman, R. G., *Dust Control and Air Cleaning*, Pergamon Press, Oxford, 1974.

Faraday Society, *Disperse Systems in Gases; Dust, Smoke and Fog*, Gurney & Jackson, London, 1936.

Fortier, A., *Mécanique des Suspensions*, Masson et Cie, Paris, 1967.

Iinoya, K., *Dust Collection*, (in Japanese), Nikkan-Kogyo-Shinbun-Sha, Tokyo, 1980.

Green, H. L. and Lane, W. R., *Particulate Clouds, Dusts, Smokes and Mists*, E. & F. N. Spon Ltd., London, 1964.

Kano, T., *Motion of the Solid-Particles*, (in Japanese), Sangyo-Gizitzu Center, Tokyo, 1977.

Ledbetter, J. O., *Air Pollution, Part B, Prevention and Control*, Marcel Dekker, New York, 1974.

Lewis, A., *Clean the Air*, McGraw-Hill, New York, 1965.

Lin, B. Y. H., *Fine Particles*, Academic Press, New York, 1976.

Mallette, F. S., *Problems and Control of Air Pollution*, Van Nostrand, New York, 1955.

Marchello, J. and Kelly, J. J., *Gas Cleaning for Air Quality Control*, Marcel Dekker, 1975.

Meldau, R., *Handbuch der Staubtechnik*, Erster and Zweiter Volumes, VDI-Verlag, Düsseldorf, 1958.

Morikawa, Y., *Two-Phase Flow of Fluid-Solid*, (in Japanese), Nikkan-Kogyo-Shinbun-Sha, Tokyo, 1980.

Ogawa, A., *Cyclone Dust Collectors*, (in Japanese), Earth Publishing, Tokyo, 1980.

Ogawa, A., *Vortex Flow*, (in Japanese), Sankaido Publishing, Tokyo, 1981.

Parker, A., *Industrial Air Pollution Handbook*, McGraw-Hill, London, 1978.

Richardson, E. G., *Aerodynamic Capture of Particles*, Pergamon Press, Oxford, 1960.

Schweitzer, P. A., *Handbook of Separation Techniques for Chemical Engineerings*, McGraw-Hill, New York, 1979.

Štorch, O., *Industrial Separators for Gas Cleaning*, Elsevier, Amsterdam, 1979.

Thring, M. W., *Air Pollution*, Butterworths, London, 1957.

Weber, E. and Broke, W., *Apparate and Verfahren der industriellen Gasvereinigung*, Vol. 1, O. Oldenbourg Verlag, Munich, 1973.

The Institution of Chemical Engineers, *A Users Guide to Dust and Fume Control*, Institute of Chemical Engineering, London, 1981.

Zimon, A. D., *Adhesion of Dust and Powder*, Plenum Press, New York, 1969.

Whytlaw-Groy, R. and Patterson, H. S., *Smoke*, Edward Arnold, London, 1932.

Gordon, G. M. and Pejsahov, I. L., Pyleuablavlivanie i Ochistka Gazov v Tsvetnoi Metallurgii, *Moskva Metallurgiia*, 1977.

Grigorjev, M. A. and Pokrovskij, G. P., *Avtomobiljnye i Traktornye Tsentrifugi*, Moskva, 1961.

Finally, I should like to thank the Managing Editor of CRC Press, Sandy Pearlman and Anita Hetzler, Coordinating Editor, for their suggestions.

These volumes could not have been completed without their kind help.

Akira Ogawa
July 1982, Tokyo

TABLE OF CONTENTS

Volume I

Chapter 5

Wet Scrubber 143

Chapter 1

INTRODUCTION

I. VARIATIONS OF THE CYCLONE DUST COLLECTOR

There are many problems concerning air and water pollution and energy dissipation all over the world. One example is the variation of cyclone dust collectors. Figure 1 shows the variation of geometries, cut-sizes, and energy-dissipations of the cyclone dust collectors.[1]

Concerning the geometry of the cyclones, denoting that a symbol D1 means an outer diameter and a symbol Ht means the total height, so the value of Ht/D1 in 1900 was nearly 0.5, that in 1930 was nearly 1.0 to 1.2, and that in 1950 was nearly 1.5 to 2.5. Then, concerning the cut-size Xc (particle size) which means that 50% of the feed dust can be separated and the remaining 50% of the feed dust cannot be separated, the cut-size Xc in 1900 was Xc ≒ 40 μm; that in 1930 was Xc ≒ 25 μm; and that in 1950 was nearly Xc ≒ 5 μm. In addition to this, symbol D.S.E. (Drehströmungsentstauber or Rotary Flow Dust Collector) was developed in Kraftwerk Union AG (1973) in West Germany.[2] The cut-size Xc of D.S.E. can easily be reached to the value Xc ≒ 0.5 μm.

On the other hand, concerning the energy-dissipation Φ (W), Φ in 1900 was nearly 400 W, Φ in 1950 was nearly 800 to 1000 W, and still more Φ of D.S.E. in 1970 was $\Phi = 10^4$ W. Therefore, decreasing the cut-size Xc from 1900 to 1970, the energy-dissipation Φ (W) was increased rapidly.

The main size (cyclone diameter D1 = 1m), the particle density ρ_p ($= 2 \times 10^3$ kg/m³), and the driving condition (inlet velocity Vo in the inlet pipe = 10 m/s) are described in Figure 1. In this example, one of the most important factors for cyclone dust collectors is that a large quantity of energy-dissipation is required to separate the sub-micron solid particles. Consequently from the energy consumption point of view, the engineer of dust collectors must select the optimal type of dust collector based upon the particle size and the particle characteristics.[3]

II. BEHAVIORS OF SOLID PARTICLES IN THE GAS FLOW

The most important fundamental and physical concept for the separations of the solid particles in the gas flow is to grasp the behavior of the motions of sub-micron particles, fine particles, and coarse particles.

Figure 2 shows the three types of flow patterns and the motions of solid particles. Figure 2(a) illustrates the flow pattern of laminar flow along a flat plate. On the same figure, the three types of behavior of solid particles in a gravitational field are shown. A coarse particle (c.p.) falls to the plate nearly linearly and a fine particle (f.p.) of the diameter Xp = 5 to 10 μm falls not linearly, but parabolically. On the other hand, a sub-micron particle (s.m.p.), smaller than diameter Xp ≒ 0.5 μm, continues to drift in the laminar gas flow and also shows a random Brownian motion due to the collisions with the molecules of gas. Therefore from the physical point of view, in the case of only a gravitational force field without an electric force, the magnetic force, or another force field, we cannot control the motion of the sub-micron solid particles in the laminar gas flow. The reader must take notice of this phenomenon. Figure 2(b) illustrates the flow pattern of the turbulent flow along the flat plate. On the same figure, the three types of the behaviors of the solid particles in a gravitational field are shown. In the case of the turbulent gas flow, there are nuisance fluctuating velocity components

$$\sqrt{\overline{v_x^2}} \,,\, \sqrt{\overline{v_y^2}} \text{ and } \sqrt{\overline{v_z^2}}$$

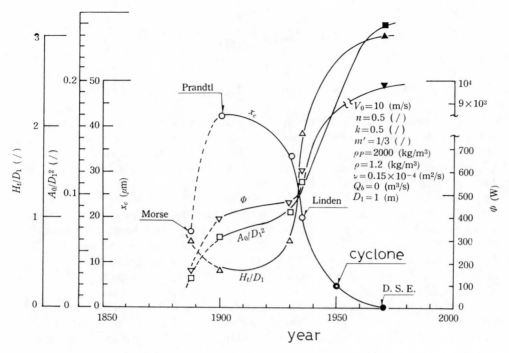

FIGURE 1.   Vicissitudes of the cyclone dust collectors.

for the x-axis, y-axis, and z-axis, respectively. Those fluctuating (turbulent) velocities hinder the separation of the fine solid particles in the separation chamber. In addition to those, the turbulent velocities contribute to create the additive pressure drop due to the Reynolds stress τ (N/m²) defined as

$$\tau_{xy} = -\rho\,\overline{v_x \cdot v_y} \qquad \tau_{yz} = -\rho\,\overline{v_y \cdot v_z} \qquad \tau_{zx} = -\rho\,\overline{v_z \cdot v_x}\,.$$

The motion of the coarse particle, which depends on the time mean velocity distribution and the turbulent velocity, falls to the plate linearly or parabolically. The fine solid particle does not always fall to the plate, but continues to follow up in the turbulent gas flow. A criterion which may be used to estimate follow-up or non-follow-up in the turbulent gas flow is the scales of turbulence,[4] namely, micro-scale or integral scale, and with the terminal velocity Wsg = $\rho_p \cdot g \cdot Xp^2/18 \cdot \eta$ of the solid particle in quiet gas. Consequently, the separation or the control of fine solid particles and of sub-micron particles in the turbulent gas flow becomes too difficult in the gravitational separation chamber.

   Then, in order to separate those fine and sub-micron solid particles in the turbulent gas flow, we need a centrifugal force, a thermal gradient force,[5,6] an electric force, or a magnetic force. From the engineering point of view, most of the gas flow in the duct, pipe, or separation chamber belongs to the turbulent flow. In the flow field, the reader must distinguish between the laminar and turbulent flows. Figure 2(c) illustrates the flow pattern of turbulent rotational flow in the confined vortex chamber such as a cyclone dust collector. In the centrifugal flow field, the fine solid particles with the tangential velocity Vθ accept the centrifugal force

$$\frac{\pi}{6} \cdot Xp^3 \cdot \rho_p \cdot \frac{V_\theta^2}{r}\Bigg)$$

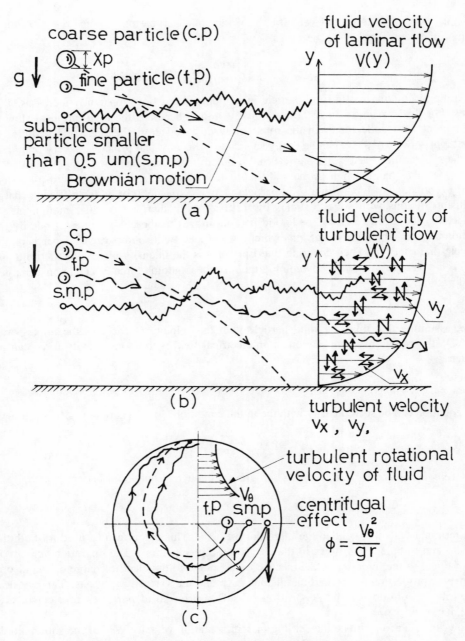

FIGURE 2.   Three types of flow patterns and of motions of solid particles.

on a radius r. Then the centrifugal effect Φ, which is defined as a ratio of the centrifugal force to the gravitational force, becomes

$$\Phi = \frac{\dfrac{\pi \cdot Xp^3}{6} \cdot \rho_p \cdot \dfrac{V_\theta^2}{r} \quad \text{(centrifugal force)}}{\dfrac{\pi \cdot Xp}{6} \cdot \rho_p \cdot g \quad \text{(gravity force)}} = \frac{V_\theta^2}{g \cdot r} \tag{1}$$

Therefore, in the centrifugal force field, the mass of the fine solid particle increases by the magnitude of the centrifugal effect Φ in comparison with the gravitational force. For example, when the tangential velocity Vθ of the solid particle on a radius r = 50 mm is 50 m/s, the centrifugal effect becomes

$$\Phi = \frac{V\overset{2}{\theta}}{g \cdot r} = \frac{2500}{9.8 \times (50/1000)} \doteqdot 5000.$$

Accordingly, the mass of a solid particle in a rotational gas flow apparently increases 5000 times. For this reason, even if the rotational gas flow becomes the high turbulent flow in the cyclone dust collector or in the rotary flow dust collector, the fine solid particles of about Xp = 0.5 to 10 μm can be easily separated. This physical principle is very important for the separation of fine solid particles in turbulent rotational gas flows.

On the other hand, when a fine solid particle is charged in an electric field by a field charge (collisions of ions with a solid particle in an electric field) or by the diffusion charge (collision of ions with solid particles by diffusion with thermal irregular motion of gas molecules), an electrically charged solid particle moves to the collecting electrode by the Coulomb force. The numbers of ions which are collided with solid particles depends on the particle diameter Xp. Then denoting that Ep(v/cm) is the intensity of the electric field and q(C) is the electric charge as shown in Figure 3, so the Coulomb force Fe acting on a charged solid particle is

$$Fe = q \cdot Ep \tag{2}$$

Here, assuming that this fine solid particle with the velocity Up accepts the Stokes drag force D, the equation of the motion in an electrical field without the gravity force can be written

$$\rho_p \frac{\pi \cdot Xp^3}{6} \cdot \frac{dUp}{dt} = q \cdot Ep - 3 \cdot \pi \cdot \eta \cdot Xp \cdot Up \tag{3}$$

This equation can be solved easily with the initial condition Up = 0 at time t = 0 as follows:

$$Up = \frac{q \cdot Ep}{3 \cdot \pi \cdot \eta \cdot Xp} \left\{ 1 - \exp\left(-\frac{18 \cdot \eta \cdot t}{\rho_p \cdot Xp^2}\right) \right\} \tag{4}$$

Therefore, when the lapse of time t(s) progresses enough, Equation 4 becomes

$$Up = \frac{q \cdot Ep}{3 \cdot \pi \cdot \eta \cdot Xp} \tag{5}$$

This terminal velocity is called the migration velocity. This theory can be used for both the laminar and turbulent gas flows in the electrostatic precipitators. The migration velocity is proportional to the intensity of the electric field Ep and to the electric charge q, and also is an inverse proportion to the particle diameter Xp and the viscosity η of gas. This physical principle is applied for the separation of sub-micron solid particles in electrostatic precipitators.[7,8]

Now, in order to estimate the settling or sedimentation velocity Wsg of the solid particle of diameter Xp in the air or in the gravitational field, the terminal velocity Wsg and the mean displacement $\sqrt{\overline{\Delta X^2}}$ per unit time by Brownian motion for the particle density ρ_p = 1 g/cm³, air temperature T = 293 K, and the pressure p = 101.3 kPa is shown in Figure 4. Here the settling or terminal velocity Wsg is the constant velocity which means that the solid particle is falling with a constant velocity in a gas or in a fluid.

In Figure 4, it can be discerned that the terminal velocity Wsg of the particle diameter Xp ≒ 0.5 μm is equal to the mean displacement $\sqrt{\overline{\Delta x^2}}$ per unit time.[9] Therefore, when we consider the motion of fine solid particles in the gravitational field, we must always remember the particle diameter Xp ≒ 0.5 μm.

Figure 5 shows the viscosity η (Pa·s) and the kinetic viscosity υ (cm²/s) of air at the pressure p = 10⁵ Pa. Also, Figure 6 shows the terminal velocity Wsg for the various kinds of the particle density ρ_p in the air of T = 288 K and of p = 101.3 kPa by Barth (1963).

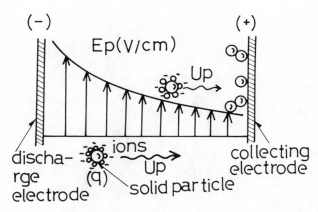

FIGURE 3. Illustration of a charged solid particle in an electric field.

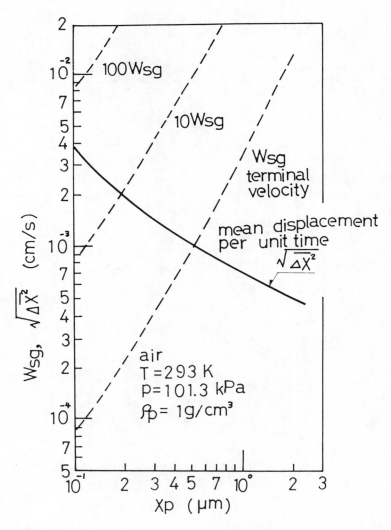

FIGURE 4. Terminal velocity and the mean displacement per unit time of fine solid particle.

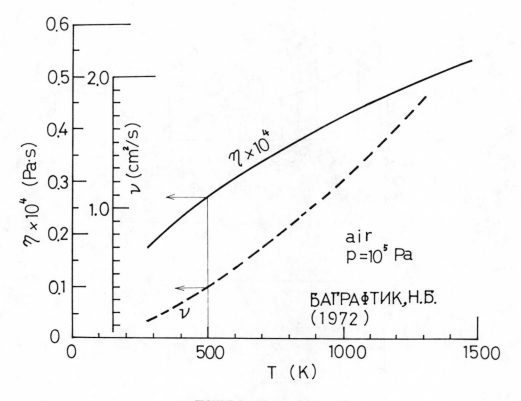

FIGURE 5.   Viscosity of air.

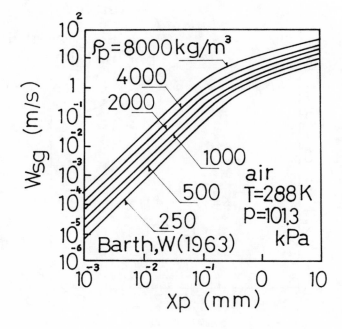

FIGURE 6.   Terminal velocity of a spherical particle in quiet air.

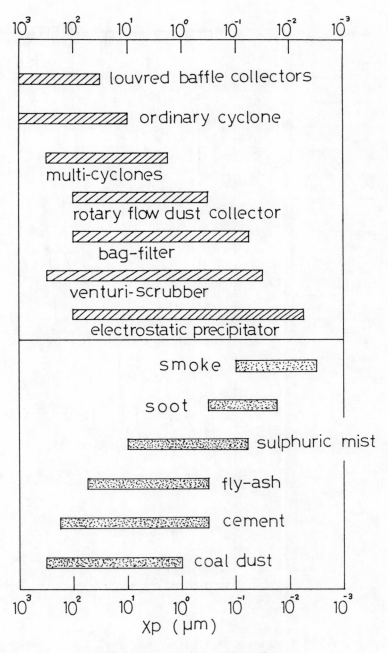

FIGURE 7. Particle sizes and dust collectors.

III. PARTICLE SIZES AND DUST COLLECTORS

Figure 7 shows an outline of the particle size boundaries for the various kinds of dust and particles and of ordinary dust collectors which can be applied to limited particle sizes. Recently, from the viewpoint of the energy utilization of coal and expansion of the traffic network, there are many dust problems. For example, a new problem of silicosis (dust disease) occurs in coal mines and in tunnel construction. Because of the development of a new type of rock drill, very fine coal dust and fine particles of stone are generated in the tunnels and in coal mines.[10]

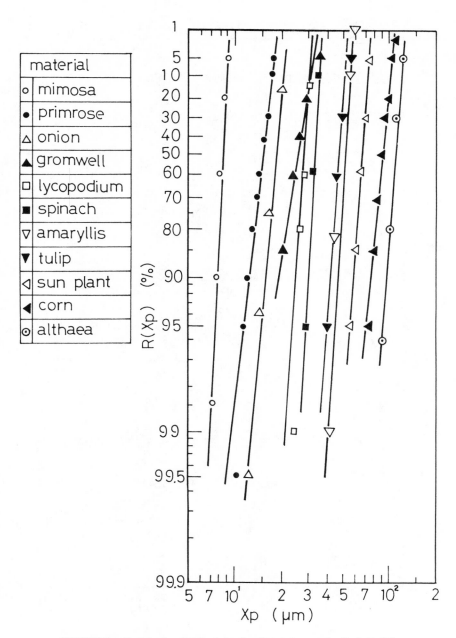

FIGURE 8.   Particle size distributions of pollen and spores by Prof. S. Miba.

It is very important to pay attention that the sizes of solid particles or dust depends on the generated location, the weather (temperature, moisture, pressure, and wind), and the mechanisms of the dust generation. Also, Miwa (1972)[11] measured very interesting particle size distributions of pollen and spores. Figure 8 shows the cumulative number residue distributions $R(Xp)$ of the particles. From this figure we can see that the distribution of the pollen and the spore particle sizes is almost the same.

IV. TYPES OF DUST COLLECTORS AND THEIR CHARACTERISTICS

Barth, an authority on cyclone dust collectors, calculated the collection efficiency by the

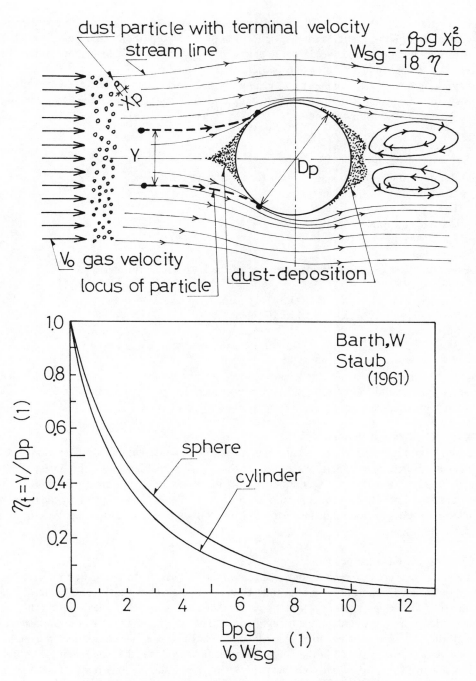

FIGURE 9.   Collection (target) efficiency and the inertia parameter.

loci of dust particles around a cylinder and around a sphere of diameter Dp in 1955.[12] One example of collection efficiency (target efficiency) η_t = Y/Dp is shown in Figure 9. In this figure, the domain of particle collision on the sphere or on the cylinder of diameter Dp is Y. Consequently, the target efficiency η_t is defined as η_t = Y/Dp. Also Wsg means the terminal velocity of a solid particle of diameter Xp in gas. On the other hand, a dimensionless quantity Wsg·Vo/g·Dp is generally called an inertia parameter.

The inertia parameter is one of the most important dimensionless quantities in considering the separation of solid particles in gas, or for motion of solid particles in a fluid flow. The physical explanation of the inertia parameter will be given in the chapter concerning the mechanical similarity law of the solid particle in the fluid flow.

In 1950, Ruhland[13] wrote a paper concerning the problems of dust collectors in the cement industry. In order to select equipment for dust separation or dust control, he found it necessary to take notice of the following items:

1. Physical property of gas
2. Gas temperature
3. Humidity and sulfuric acid content for dust-laden gas
4. Concentration of dust-laden gas
5. The sort of dust
6. Density and form of dust
7. The distribution of particle size

Generally, West Germans divided the size of the particles as follows:

Coarse particle	$X_p = 100$ to 1000 μm
Fine particle	$X_p = 10$ to 100 μm
Finest particle	$X_p = 0.1$ to 10 μm
Smoke	$X_p = 0.001$ to 0.1 μm

Also, the main equipment for dust separation or for dust control, which were set in mills, furnaces, and general industry, could be briefly described for gas treatment at 20,000 m³/hr in the following sections.

A. Rotex Centrifugal Separator

As shown in Figure 10, the dust-laden gas flows through the rotating cylinders and rotates by the wall friction on cylinders. As a result of this rotation, the dust (solid) particles, which accepted the centrifugal force, accumulate on the inner surfaces of the rotating cylinders by centrifugal force. Therefore, the separation of dust or solid particles is realized by the relative velocity between gas flow and dust particle flow. This equipment can be applied to high temperature gas.

B. Cyclone Dust Collectors

There are two types of cyclone dust collectors: the axial inlet flow (uni-flow) cyclone as shown in Figure 11 and the tangential inlet flow cyclone as shown in Figure 12. In Figure 11, the dust laden gas in the cyclone body is rotated by the guide vanes and the dust or solid particles are separated by centrifugal force. The clean gas flows to the atmosphere through the exit pipe or inner pipe. This type of axial cyclone can be applied in a multi-cyclone system. In Figure 12 in the tangential inlet flow cyclone, the dust-laden gas rotates by flowing from the tangentially connected inlet pipe into the cyclone body. This type of tangential flow cyclone can be applied as a multi-cyclone system and also can be used for high temperature gas. From a practical point of view, solid particles larger than $X_p = 13$ μm can be easily separated.

C. Bag Filter

The bag filter as shown in Figure 13 can be used for nearly perfect purification of dust-laden gas. Also in the case of the dense filter (textile), the fines do not depend on the separation of dust. The bag filter is sluggish for fluctuating gas flow rates, but always

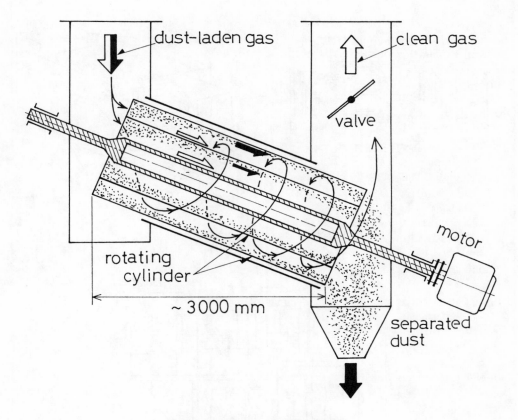

FIGURE 10.   Rotex centrifugal separator.

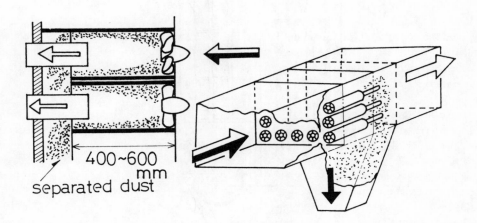

FIGURE 11.   Axial flow cyclones.

accompanies the high wear and tear expenses for high dust-laden gas flow. Besides treating high humidity gas, gas-containing SO_3, or gas temperatures higher than $T = 373$ K, application of the bag filter may be difficult.

D. Electrostatic Precipitator

The electrostatic precipitator as shown in Figure 14 needs very wide space due to the slow velocity ($\leqq 2$ m/s) of dust-laden gas in the precipitator. When the velocity of the dust-

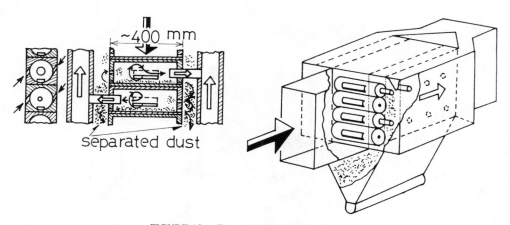

FIGURE 12.   Tangential inlet flow cyclones.

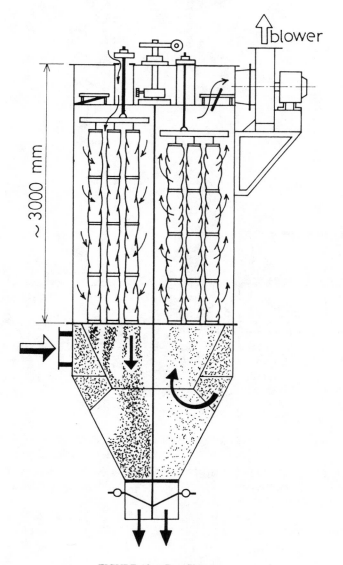

FIGURE 13.   Bag filter.

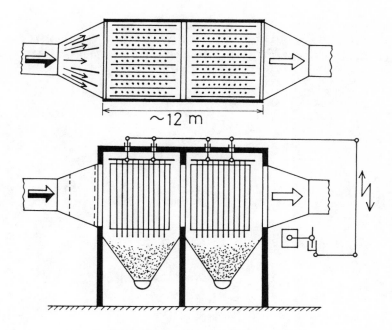

FIGURE 14.   Electrostatic precipitator.

laden gas becomes slower, the separation efficiency becomes better due to the increment of the residue time in the precipitator. This characteristic is the same as the bag filter. Then the electrostatic precipitator can be used for gas temperatures approaching the dew point, but for gas temperature higher than T = 523 K or more and for high content vapor, the application of electrostatic precipitators becomes difficult. Under most reasonable driving conditions, it is possible to separate the dust of Xp = 0 to 10 μm.

Summarizing the application of dust separators to the cement production industry, the system of dust collection (separator) can be estimated for the treatment of flow rate of dust-laden gas Qo = 100,000 m³/hr. This quantity of gas corresponds to the production of 350,000 kg/day by a rotating furnace. The required space contains not only separation equipment, but also the ventilation equipment accompanying the pipe lines. The power was estimated under the condition of gas temperature T = 393 K. The estimated results for the various dust separators are tabulated in Table 1.

E. Fractional Collection Efficiency

Anselm (1950) described the characteristics of the separation for the various kinds of the dust collectors.[14,18] In order to indicate the separation grade for the particle size Xp, we always use the curve of the fractional collection efficiency η_x (Xp). Here we define the cut-size Xc50 which is the particle size corresponding to the fractional collection efficiency η_x (Xp) = 0.5 (50%).

Figure 15 shows the experimental curves of the fractional collection efficiencies η_x (Xp) for gravitational sedimentation, cyclones, and the electrostatic precipitators. In the case of cyclones, where the cyclone diameter D1 is decreased from D1 = 3000 mm to D1 = 100 mm, the cut-size Xc50 is also decreased from Xc50 = 35 μm to Xc50 = 1.5 to 2 μm. Here we must pay attention that the cut-size Xc50 always depends on the driving conditions (for example, the inlet velocity Vo of gas in the inlet pipe of cyclones) of fluid flow and the physical characteristics of dust in the separators. Table 2 shows the general numerical values of the dust collectors.

Table 1
COMPARISON OF SEPARATION SYSTEMS UNDER DRIVING
CONDITIONS WITH GAS FLOW RATE Qo = 100,000 M³/HR AND
GAS TEMPERATURE T = 393 K

	Dim.	Rotex	Centrifugal separator	Multi-cyclone	Bag filter	Electrostatic precipitator
Required space	m	900	600	300	500	1100
Maximum approved gas temperature	K	630	680	680	370	470
Cut-size Xc	μm	—	10—25	—	0.5—10	0.5—10
Approved particle size Xp	μm	1—1000	13—1000	1—100	0.1—100	0.1—100
Separation efficiency η_c for above stated particle size	%	98		95	99.7	99.4
Power	kw	34	110	106	110	29

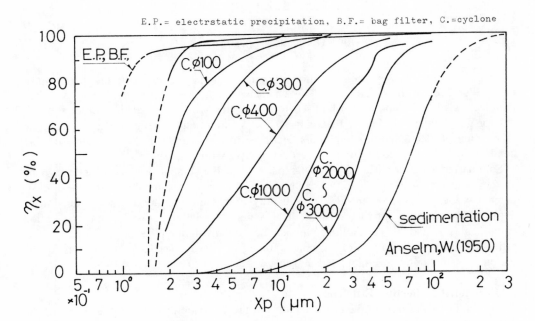

E.P.= electrstatic precipitation, B.F.= bag filter, C.=cyclone

FIGURE 15.   Curves of fractional collection efficiencies.

V. TYPES OF DUST COLLECTORS

Figure 16 shows the types of the settling chambers.[15] This type of separator is applied to the separation of coarse solid particles.

This type of the dust bag separator, as shown in Figure 17, is adapted as a preliminary air purification for very high dust concentrations. The dust-laden gas having velocity Vo in an inlet pipe enters into a separation space of a cylinder D1. When this dust laden gas returns to 3.14 rad with a center of curvature r, the solid particles are separated from the air flow by centrifugal force. One numerical example is as follows: D1 ≒ 10 m, L ≒ D1 ≒ 10 m, Vo = 10 m/s, V1 ≒ 1 m/s, Xp ⩾ 25 to 30 μm, η_c = 65 to 85%, Δp_c = 150 to 390 Pa.

15

Table 2
GENERAL NUMERICAL VALUES OF DUST COLLECTORS[14]

Type	Xc (μm)	Vo (m/s)	Δp (Pa)	Specific energy dissipation (kW/1000m³·h·K)	T (K)	Required minimum space (m³/1000 m³)
Sedimentation	35	0.5—0.8	10—15	0.01—0.015	773	10—12
Special sedimentation	35	1.0—2.0	20—40	0.02—0.04	773	8—10
Cyclone D_1 = 2—3 m	25	12.0—18.0	500—1000	0.25—0.50	673	0.5—1.0
Cyclone D_1 = 0.4—1 m	10—16	—	400—1200	0.25—0.75	673	2.0—5.0
Cyclone D_1 = 0.1—0.4m	3.5—60	—	400—1200	0.25—0.75	673	0.5—4.0
Cyclone D_1 = 0.1 m	2.5	—	—	0.05—0.20	673	0.1
Bag filter	0.5—1.0	—	500—1500	0.3—1.1	373	2.7—5.0
Electrostatic precipitator	0.5—1.0	0.5—1.0	30—100	0.1—0.3	473—573	8—12

Δp: Pressure drop at T = 293 K.
T: Gas temperature.

In this figure, the inlet pipe shown with a dotted line has a diffused pipe with θ = 0.21 rad. In order to decrease the velocity Vo of a dust-laden gas flow in the separation space and to increase separation efficiency η_c, this diffused pipe is sometimes applied.

Figure 18 shows the separation principle of the reflection inertia dust collector.

Figure 19 shows a louver dust separator. The separation efficiency (collection) η_c (%) depends on the shape of the louver and on the ratio of the flow rates between the clean gas and the cyclone.

Figure 20 shows the types of baffle chambers. These types of separators are applied for the separation of coarse dusts or coarse solid particles.

Figure 21 shows the types of cyclone dust collectors.

Figure 22 shows multi-cyclones with guide vanes and with a common dust hopper (bunker).

Figure 23 shows multi-cyclones with a tangential inflow and with a common dust bunker.

Figure 24 shows multi-cyclones with a separately dust-laden gas inflow.

Figure 25 shows one type of bag filter.

Figure 26 shows one type of bag filter.

Figure 27 shows one type of bag filter. For these bag-filter systems, there are several technical methods of de-dusting the surface of a bag filter.

Figure 28 shows an electrostatic precipitator of a simple stage tubular type. As shown, the collector electrode (collecting electrode) connected to the earth has cylindrical and hexagon shapes.

Figure 29 shows an electrostatic precipitator with a plate type collecting electrode.

Figure 30 shows the simple spray chamber.

Figure 31 shows a packed bed scrubber.

Figure 32 shows a scrubber with baffles.

VI. COLLECTOR SYSTEM

A. Production Plant of Carbon Black

Figure 33 shows one example of a production plant of carbon black (mean diameter Xp

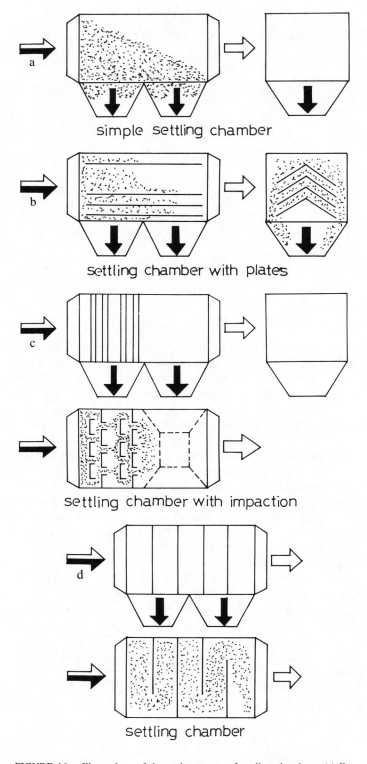

FIGURE 16.   Illustrations of the various types of settling chambers. (a) Illustration of a simple settling chamber. (b) Illustration of a settling chamber
with plates. (c) Illustration of a settling chamber with plate. (d) Illustration of
a settling chamber.

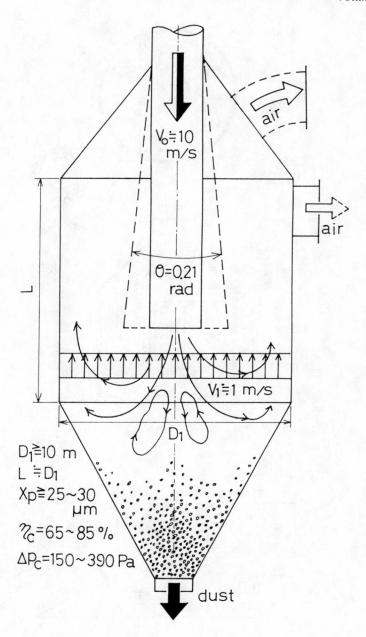

FIGURE 17.   Dust bag.

$\doteq 0.03$ μm) of 50,000 kg/day in 1955.[16] In this plant the electrostatic precipitator was applied to agglomerate the carbon-black particles. The agglomerated particle diameter became Xp = 0.4 to 10 μm. The two-stage large cyclone dust collectors were connected and then the bag filters of 4000 bags (diameter 127 mm, length 3048 mm) were set. The filtration velocity Vo of the bag filter kept about Vo $\doteq$ 6.5 m/s.

B. Air Cleaner for a Bus

Figure 34 shows an air cleaner system of the engine of a large bus.[17] The dust-laden gas enters the axial flow cyclone and the fine dust escaping from the cyclone is accepted by the

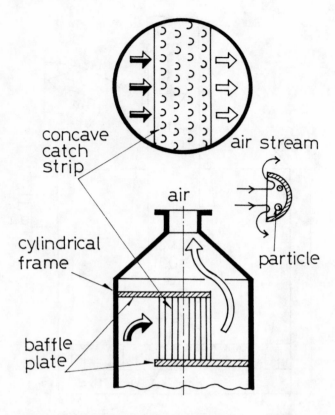

FIGURE 18. Separation of reflection inertia dust collector.

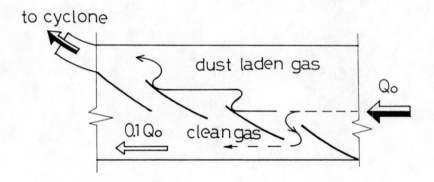

FIGURE 19. Louver dust collector.

centrifugal force just in front of the entrance of the filter. Some parts of the dust are caught on the surface of the oil. The clean gas flows through the filter flows into the engine.

VII. FLOW DIAGRAM FOR DESIGN OF A DUST COLLECTOR SYSTEM

In order to design a dust collector system[3] it is convenient to consider the flow chart, shown in Figure 35. First of all, we must estimate the distribution of the particle size R (Xp) and the physical characteristics of the particle or dust. Then we must consider what

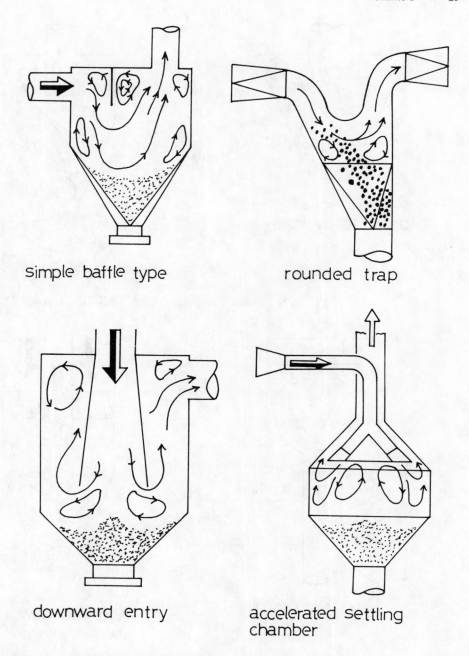

simple baffle type

rounded trap

downward entry

accelerated settling
chamber

FIGURE 20.   Illustration of the various kinds of baffle chambers.

kinds of the operating conditions, including the flow rate Qo (m³/hr) of gas, the gas temperature T (K), and the total pressure drop Δp_c (Pa). After that, we must design or select
what kind of dust collectors, the dimensions, the space, and the maintenance required, for
the optimum condition.

In the next step, after selecting the dust collector, we must consider the fractional collection
efficiency $\eta_x(Xp)$ and the cut size Xc50. Also, we must estimate the required power for the
total system, including pipes, ducts, power for the total system, including pipes, ducts,

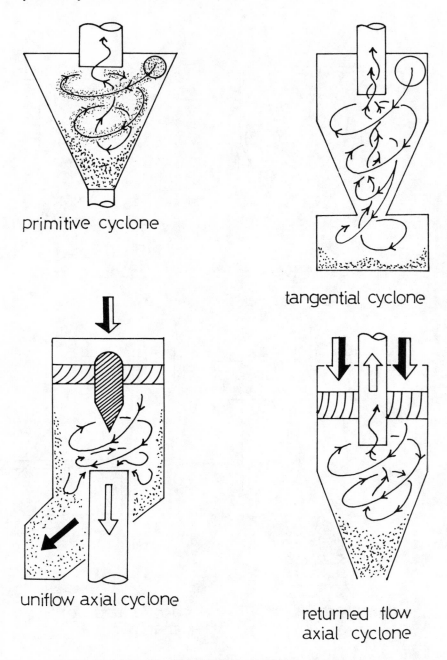

primitive cyclone

tangential cyclone

uniflow axial cyclone

returned flow
axial cyclone

FIGURE 21.   Illustration of various kinds of cyclone dust collectors.

separating systems, etc. After that we must consider the reliability and maintainability of
the total system, including the dust collector. By this estimation, we may guarantee exactly
how many years the total system can be operated.

Next, we must estimate the total collection efficiency η_c (%), the total cost of the dust
collectors, the auxiliary equipment, and the availability of the total system. Lastly, we must
conclude the final judgment for the total system, including the dust collectors. On the other
hand, if this system does not satisfy the optimum condition, we must again try to estimate
from the first step.

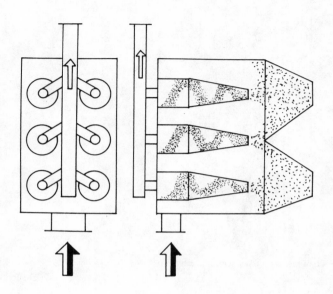

FIGURE 23.   Illustration of the multi-cyclones with tangential inflow and with a common dust bunker.

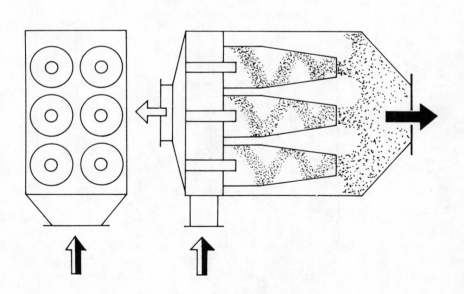

FIGURE 22.   Illustration of multi-cyclone dust collectors with guide vanes and with common dust hopper (bunker).

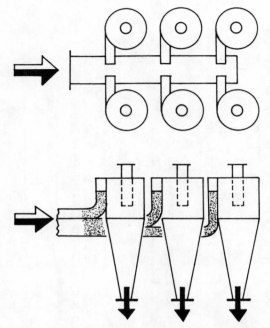

FIGURE 24.   Illustration of multi-cyclones with separately dust-laden gas inflow.

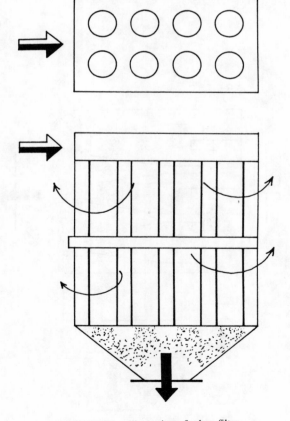

FIGURE 25.   Illustration of a bag filter.

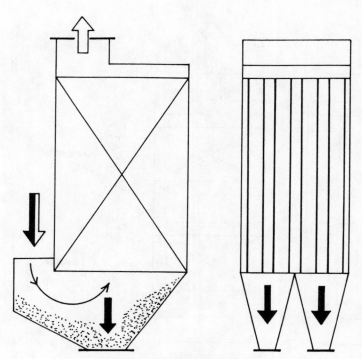

FIGURE 26.   Illustration of a bag filter.

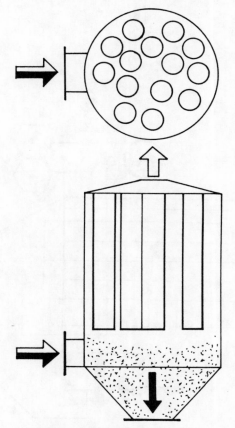

FIGURE 27.   Illustration of a bag filter.

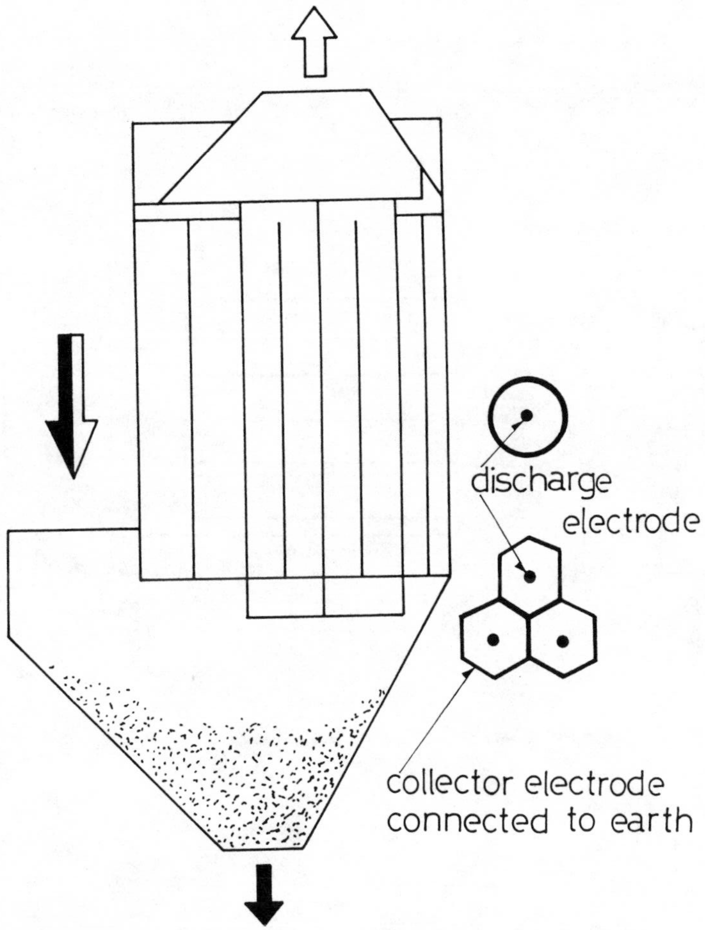

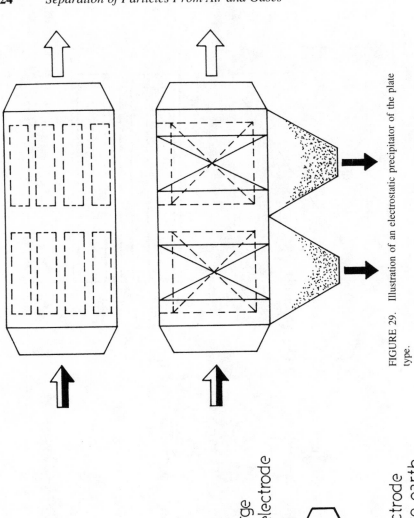

FIGURE 29. Illustration of an electrostatic precipitator of the plate type.

discharge electrode

collector electrode connected to earth

FIGURE 28. Illustration of an electrostatic precipitator (single stage tubular).

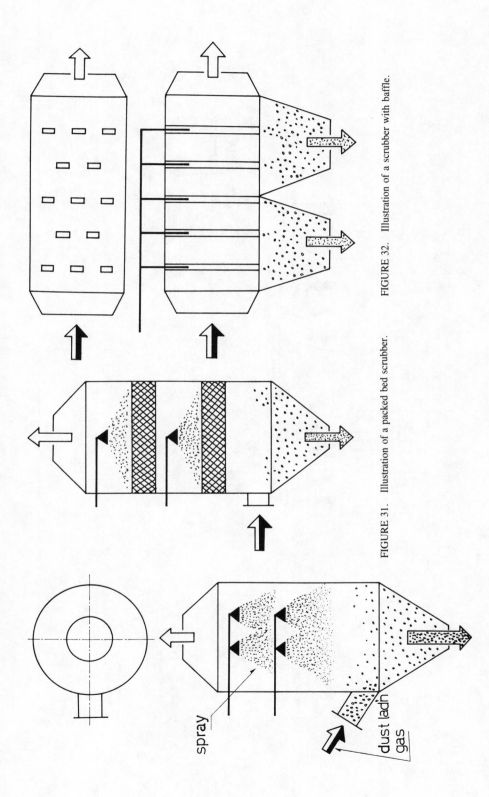

FIGURE 32. Illustration of a scrubber with baffle.

FIGURE 31. Illustration of a packed bed scrubber.

spray

dust ladn
gas

FIGURE 30. Illustration of a simple spray chamber.

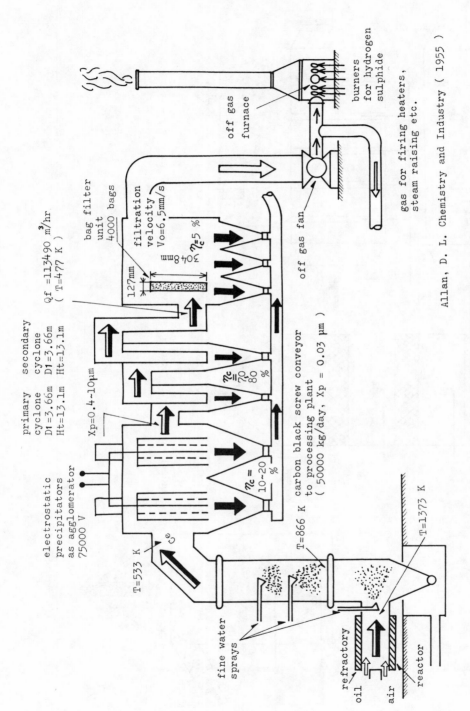

FIGURE 33.   Diagrammatic layout of a collection system for carbon black.

Allan, D. L. Chemistry and Industry (1955)

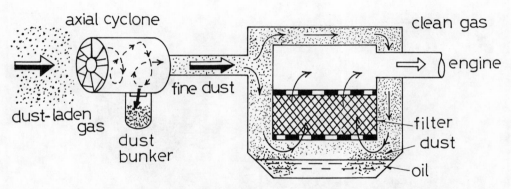

FIGURE 34.   Illustration of an air filtration of system for a bus engine.

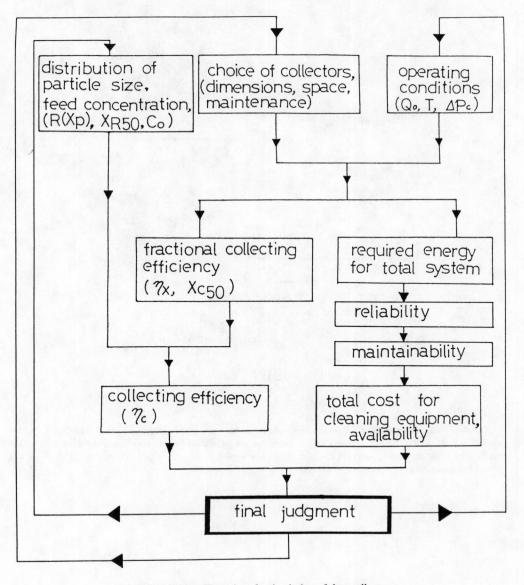

FIGURE 35.   Flow chart for the design of dust collectors.

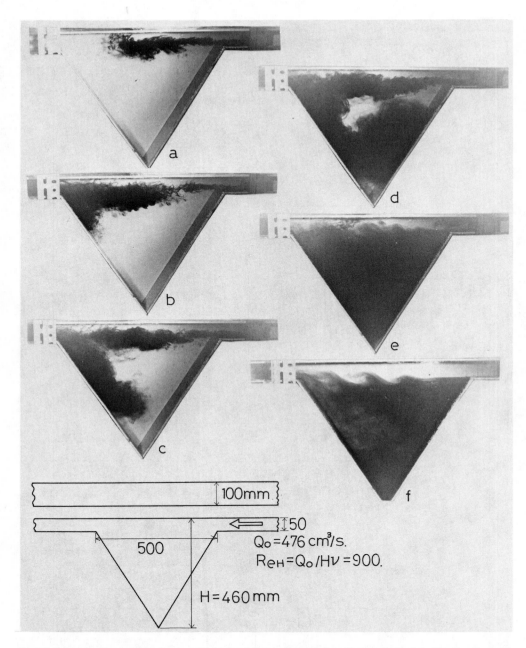

Plate 1 shows the flow pattern of water with lapse of time in the triangle sedimentation bunker. The flow Reynolds number is ReH = 900. From this plate, it can be seen that the circulating flow in the triangle sedimentation bunker has occurred and the vortex street on the high shear flow region is also developed. These flow phenomena will hinder the sedimentation separation of the solid particles in the dust bunker.

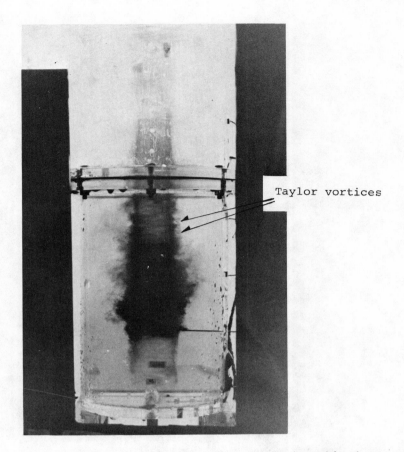

Taylor vortices

Plate 2 shows the flow pattern of Taylor vortices surrounding the quasi-forced vortex flow of water in the vortex chamber. Diameter of a vortex chamber D1 = 140 mm. Diameter of an inner pipe D2 = 47.5 mm. Total height, Ht = 372 mm. The flow Reynolds number is Rec = 2344. It is very important to notice that Taylor vortices play an important role in the transportation of the flux of the angular momentum for the rotating fluid.

Taylor vortices
in a cone cyclone

Plate 3 also shows the flow pattern of Taylor vortices of water in a cone type of the cyclone dust collector. Flow Reynolds number is Rec = 365. Diameter D1 = 100 mm. Inner pipe D2 = 39 mm. Height Ht = 150 mm.

Plate 4 shows the behaviors of the re-entrainment for the collected dust particles by the turbulent rotational flow near the dust boundary layer in the dust bunker of the cyclone dust collector. It is very important to notice that the collection efficiency is decreased when increasing the re-entrainment. Cyclone diameter $D1 = 100$ mm. Diameter of a dust bunker $D3 = 100$ mm. Flow Reynolds number $Rcy = 4300$.

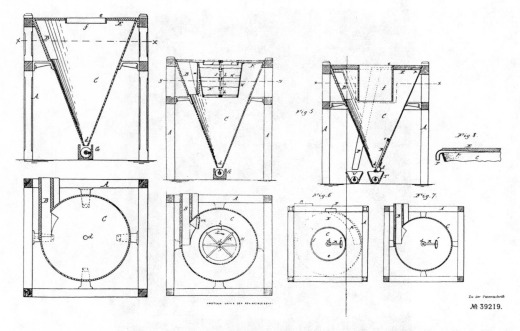

Plate 5 is a patent by the Knickerbocker Company in Jackson in 1886. This cone type of the cyclone dust collector is the most initial type. This copy (KAISERLICHES PATENTAMT) is based upon Landes-Gewerbemuseum in Stuttgart, West Germany.

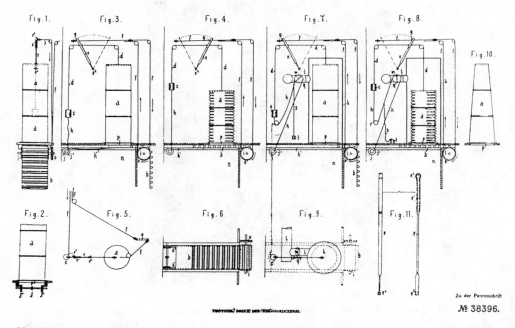

Plate 6 is a patent of Wilhelm Friedrich Ludwig Beth from 1886. Bag filter of cornice type system is the most initial type. This copy is based upon Landes-Gewerbemuseum in Stuttgart, West Germany.

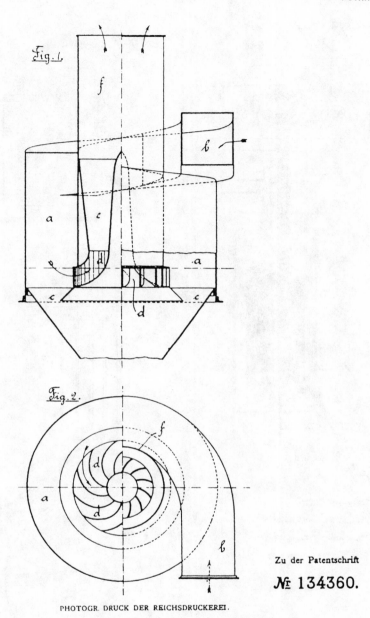

Fig. 1.

Fig. 2.

Zu der Patentschrift

№ 134360.

PHOTOGR. DRUCK DER REICHSDRUCKEREI.

Plate 7 is a patent of Vereinigte Maschinenfabrik Augsburg and Maschinen-baugesellschaft Nürnberg, A. G., from 1901. This cyclone dust collector which is set up in the guide vanes in the exit pipe is intended for converting the rotational flow into the straight flow for decreasing the pressure drop. This copy is based upon Landes-Gewerbemuseum in Stuttgart, West Germany.

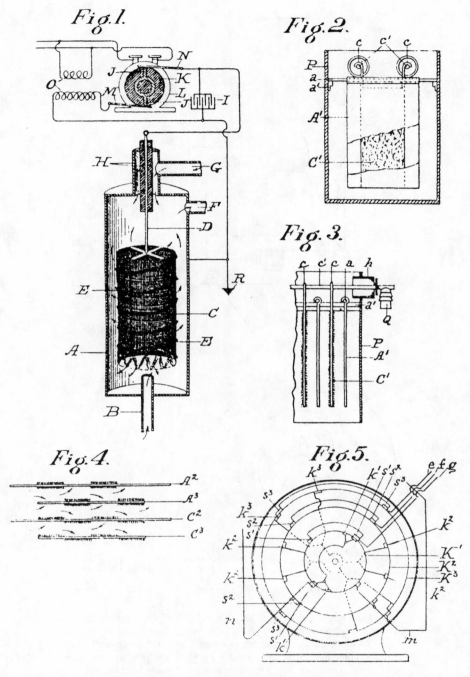

Fig.1.

Fig.2.

Fig.3.

Fig.4.

Fig.5.

Plate 8 is a patent of Dr. Frederick Gardner Cottrell from 1908. This system is the initial type of the electric precipitator. This copy is based upon Landes-Gewerbemuseum in Stuttgart, West Germany.

Zu der Patentschrift 290146

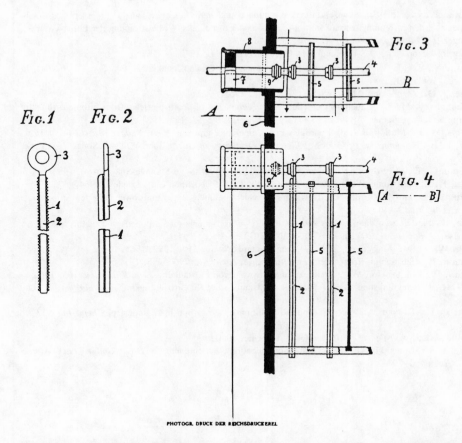

Fig. 1   *Fig. 2*   *Fig. 3*

Fig. 4
[A — — B]

PHOTOGR. DRUCK DER REICHSDRUCKEREI

Plate 9 is a patent of Erwin Möller from 1912. In this plate, Figure 1 and Figure 2 are the discharge electrodes and number 5 in Figure 3 and Figure 4 shows the collecting electrode. Number 6 in these figures shows the outer wall of the separation chamber. This copy is based upon Landes Gewerbemuseum in Stuttgart, West Germany.

REFERENCES

1. **Ogawa, A.,** Über die Staubkonzentrationsverteilungen und den Gleichgewichtsdurchmessers des Staub-teilchen und den Strömungs-verlauf in drei-dimensionaler turbulenter Wirbelströmung der Luft im Wirbel-erzeuger, *J. Coll. Eng. Nihon Univ.,* 21A, 133, 1980.
2. Kraftwerk Union AG., D.S.E. Handbuch, Erlangen, B. R. D., 1972.
3. **Ogawa, A.,** Cyclone dust collectors for general industry in Czechoslovakia, *J. Jpn. Soc. Air Pollut.,* in Japanese, 16(1), 59, 1981.
4. **Hinze, J. O.,** *Turbulence,* McGraw-Hill, New York, 1955.
5. **El-Shoboksky, M. S.,** A method for reducing the deposition of small particles from turbulent fluid by creating a thermal gradient at the surface, *Can. J. Chem. Eng.,* 59, 155, 1981.
6. **Phillips, W. F.,** Motion of aerosol particles in a temperature gradient, *Phys. Fluids,* 18(2), 144, 1975.
7. **Loeb, L. B.,** *Fundamental Processes of Electrical Discharge in Gases,* John Wiley & Sons, New York, 1939.
8. **Matzumoto, T.,** *Electrostatic Precipitator,* (in Japanese), Nikankogyo-Shinbun-Sha, Tokyo, 1980.
9. **Green, H. L. and Lane, W. R.,** *Particulate Clouds,* E. & F. N. Spon Ltd., London, 1964.
10. **Harris, W. B. and Eisenbud, M.,** Dust sampler which simulates upper and lower lung deposition, *Ind. Hyg. Occup. Med.,* 446.
11. **Miwa, S.,** Particle size distributions and the density measurements of the pollens and of the spores, (in Japanese), *J. Soc. Powder Tech. Jpn.,* 9(2), 102, 1972.
12. **Barth, W.,** Entwicklungslinien der Entstaubungstechnik, *Staub,* 21(9), 382, 1961.
13. **Ruhland, E.,** Entstaubungsfragen der Zementindustrie, *Zement-Kalk-Gibs,* 5, 1950.
14. **Anselm, W. and Anselm, W.,** Kennlinien und Auswahl von Entstaubern, *Zem. Kalk Gips,* 165, 1950.
15. **Nagel, R.,** Entwicklungstand der mechanischen Entstauber und Klassifizierungsfragen, *Staub,* 21(9), 406, 1961.
16. **Allan, D. L.,** The prevention of atmospheric pollution in the carbon black industry, *Chem. Ind.,* 1320, 1955.
17. **Glybin, A. I.,** *Avtotraktornye Filjtry,* Mashinostroenie, 1980.
18. **Meloy, T. M.,** Mineral processing separation optimization by circuit analysis, *Proc. Rindge Coal Cleaning Conf.,* August, 1979.

Chapter 2

FLUID FLOW IN DUCTS AND FLOW PAST A SPHERE AND A CYLINDER

I. INTRODUCTION

As shown in Figure 1, the pressure drop Δp (Pa) for a length L of circular pipe on the fully developed velocity flow can be described as

$$\Delta p = \lambda \cdot \frac{L}{D} \cdot \frac{\rho \overline{V}^2}{2} \tag{1}$$

where a dimensionless number λ (1) is a pipe friction factor, D(m) is a diameter of a pipe, L (m) is the length of a pipe, ρ (kg/m³) is the density of fluid flow, and $\overline{V}$ (m/s) is the mean fluid velocity in a pipe. Here the friction factor λ depends on the flow Reynolds number $Re = \overline{V} \cdot D / \upsilon$ and the surface roughness ϵ (m) of a pipe.[1,2] For the developed turbulent flow of $Re \geq 6000$ in the smooth circular pipe, Blasius (1913) derived the famous empirical equation defined as

$$\lambda = 0.316 \, Re^{-0.25} \tag{2}$$

On the other hand, for the low Reynolds number of $Re \leq 2300$, Hagen (1839) and Poiseuille (1841) derived an equation of the friction factor λ in the laminar flow for the fully developed velocity profile

$$\lambda = \frac{64}{Re} \tag{3}$$

Now in order to explain the more detailed physical meaning of the pressure drop, we consider the small distance ΔX of pipe diameter D on the fully developed flow region as shown in Figure 1. Assuming that the fluid flow is steady, the force components acting on the small fluid element of a volume $\pi r^2 \cdot \Delta X$ are as follows:

$\pi r^2 \cdot p$: Pressure force (N) on the side area at X
$\pi r^2 \cdot (p + \frac{dp}{dx} \Delta X)$: Pressure force (N) on side area at $X + \Delta X$
$2\pi r \cdot \Delta X \cdot \tau$: Frictional force (N) by the shear stress τ on the peripheral surface
of a cylinder of radius r

Then the equation of motion can be written

$$\pi r^2 \cdot \Delta X \cdot \rho \cdot \frac{dVm}{dt} = \pi r^2 \cdot p \; - \pi r^2 \cdot \left(p + \frac{dp}{dX} \cdot \Delta X \right) - 2\pi r \cdot \Delta X \cdot \tau \tag{4}$$

where Vm is the mean axial velocity of a small fluid element. From the assumption of the steady flow, dVm/dt becomes zero. Then a relationship between p and τ can be obtained from Equation 4 as

$$\tau = -\frac{1}{2} \cdot \frac{dp}{dx} \cdot r \tag{5}$$

Therefore the shear stress τ becomes the maximum value at the wall (D/2) defined as

$$\tau_m = \frac{1}{2} \cdot \frac{dp}{dX} \cdot \frac{D}{2} = -\frac{1}{4} \cdot \frac{dp}{dX} \cdot D \tag{6}$$

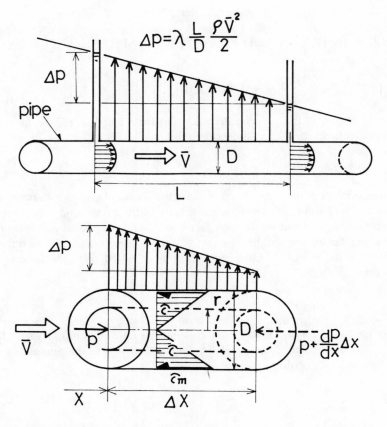

FIGURE 1.   Illustration of the pressure drop for the fully developed velocity flow in a pipe.

and to zero at the center of pipe (r = 0).

Now assume that the shear stress τ_m may be written as

$$\tau_m = C \cdot \frac{\rho \bar{V}^2}{2} \qquad (7)$$

where c is an constant value. Combining Equations 5 and 7 we can obtain the following equation

$$-\frac{dp}{dX} = 4C \cdot \frac{1}{D} \cdot \frac{\rho \bar{V}^2}{2} \qquad (8)$$

Here (dp/dx) is a decreasing value along the x-axis, so the value of (dp/dx) is negative value. Consequently, substituting L for dX, Δp for ($-dp$) and λ for 4C, we can obtain the equation of the pressure drop

$$\Delta p = \lambda \frac{L}{D} \cdot \frac{\rho \bar{V}^2}{2} \qquad (9)$$

This equation is the same as Equation 1 and is called the Darcy-Weisbach equation.

On the other hand, in the case of the laminar flow condition in pipe, the shear stress τ can be represented as

$$\tau = \eta \cdot \frac{dV}{dr} \qquad (10)$$

where V (m/s) is the axial velocity of fluid at the radius r and η (Pa·s) is the viscosity of fluid. Substituting Equation 10 into Equation 5, we can obtain the following equation

$$dV = -\frac{1}{2} \cdot \frac{1}{\eta} \cdot \frac{dp}{dX} \cdot r \cdot dr \tag{11}$$

So, integrating the above equation with the boundary condition $V = 0$ at $r = D/2$, the following equation can be obtained

$$V = -\frac{1}{4 \cdot \eta} \left(\frac{D}{2}\right)^2 \cdot \frac{dp}{dX} \cdot \left\{ 1 - \left(\frac{2r}{D}\right)^2 \right\} \tag{12}$$

Then, denoting that the center velocity at $r = 0$ is Vo, the equation of Vo becomes

$$Vo = -\frac{1}{4\eta} \cdot \left(\frac{D}{2}\right)^2 \cdot \frac{dp}{dX} \tag{13}$$

Finally, Equation 13 transforms to

$$V = Vo \cdot \left\{ 1 - \left(\frac{2 \cdot r}{D}\right)^2 \right\} \tag{14}$$

This velocity distribution shows a parabola form.

While the pressure drop for the length X denotes Δp, substituting $\Delta p/X$ into $|dp/dX|$, and also Qo (m³/s) is the flow rate, $\overline{V}$ (m/s) is the mean axial velocity in pipe, using Equations 13 and 14, we can obtain the following equation

$$Qo = \int_0^{D/2} 2 \cdot \pi \cdot r \cdot dr \cdot V = \frac{\pi \cdot \Delta p \cdot (D/2)^4}{8 \cdot \eta \cdot L} = \frac{\pi \cdot \Delta p}{8 \cdot \eta \cdot L} \cdot \left(\frac{D}{2}\right)^4 \tag{15}$$

The above equation can be transformed to

$$\overline{V} = \frac{4 \cdot Qo}{\pi \cdot D^2} = \frac{\Delta p}{8 \cdot \eta \cdot L} \cdot \left(\frac{D}{2}\right)^2 = \frac{Vo}{2} \tag{16}$$

Consequently in the case of a laminar flow, the flow rate Qo is proportional to the pressure drop Δp and also the mean velocity $\overline{V}$ is equal to Vo/2. This equation is called the Hagen-Poiseuille law. This law is applied for the estimation of the pressure drop in a porous media and a bag filter.

Then transforming Equation 16 to

$$\Delta p = 8 \cdot \eta \cdot L \cdot \left(\frac{2}{D}\right)^2 \cdot \overline{V} = \frac{32 \cdot \eta \cdot L}{D^2} \cdot \overline{V} = \frac{64}{Re} \cdot \frac{L}{D} \cdot \frac{\rho \cdot \overline{V}^2}{2} \tag{17}$$

the pipe friction factor λ becomes

$$\lambda = \frac{64}{Re} \text{ (laminar flow)} \tag{18}$$

II. FLOW IN THE INLET (ENTRANCE) LENGTH

As shown in Figure 2, the flow all over the cross-section of pipe at the inlet position has the uniform velocity distribution which is called a potential flow. Flowing downstream in the pipe, the thickness of the boundary layer (vorticity region) is increased and, at the distance L from the entrance place, the thickness of the boundary layer develops to the pipe

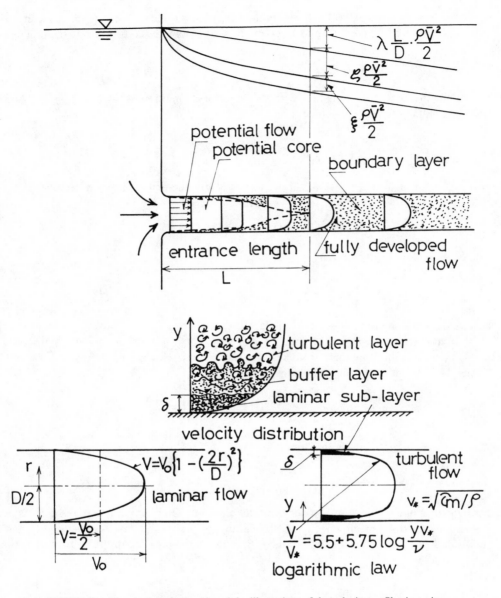

FIGURE 2.   Flow in the inlet length and the illustrations of the velocity profiles in a pipe.

center. Then flowing downstream from this position, the velocity distribution, the pipe friction factor λ, the rate of pressure drop, and the kinetic energy of the fluid in a pipe becomes a constant value. This length L is called an inlet length or an entrance length.

The following results were published concerning the entrance length L:

In the case of the laminar flow

Boussinesq (1891)	L = 0.065·Re·D	(calculation)
Nikuradse (1922)	L = 0.065·Re·D	(experiment)
Schiller (1922)	L = 0.029·Re·D	(calculation)
Asao, Iwanami, and Mori (1952)	L = 0.06·Re·D	(experiment)

In the case of the turbulent flow

Latzko (1921)	L = 20·D	(calculation)
Kirstein and Nikuradse (1927)	L = (25 to 40)·D	(experiment)

Then, the kinetic energy of fluid $\zeta\rho\overline{V}^2/2$ per unit mass of fluid flowing through the cross-section of the pipe diameter D can be represented as

$$\zeta\cdot\rho\cdot\frac{\overline{V}^2}{2} = \frac{1}{Qo}\int_0^{D/2}(2\cdot\pi\cdot r\cdot dr\cdot V)\frac{\rho V^2}{2}$$

where V (m/s) is the axial velocity of the fluid at radius r and Qo (m³/s) is the flow rate of the fluid. The value of ζ becomes $\zeta = 2$ (theoretical) for the laminar flow and $\zeta \fallingdotseq 1.09$ for the turbulent flow. In the entrance region, the thickness of the boundary layer is very thin, so the value of the velocity gradient (dV/dy) becomes very high in comparison with that in the fully developed region. Accordingly, the additive frictional pressure drop $\S\cdot\rho\cdot\overline{V}^2/2$, in addition to $\lambda\cdot(L/D)\cdot\rho\cdot\overline{V}^2/2$, is created. The values of $\S$ are determined by the various researchers as:

In the case of laminar flow

Hagen	$\S = 0.7$	(experiment)
Boussinesq	$\S = 0.24$	(calculation)
Schiller	$\S = 0.16$	(calculation)
Atkinson and Goldstein	$\S = 0.41$	(calculation)

In the case of turbulent flow

Hagen	$\S = 0.31$	(without inlet bell mouth)

Consequently the total pressure drop Δp for the downstream from the entrance length can be written

$$\Delta p = \lambda\cdot\frac{L}{D}\cdot\frac{\rho\overline{V}^2}{2} + \zeta\cdot\frac{\rho\overline{V}^2}{2} + \S\cdot\frac{\rho\overline{V}^2}{2} \qquad (19)$$

Figure 3 shows Nikuradse experiments of the velocity distribution in the smooth pipe for the various Reynolds number Re.

III. FRICTION FACTOR FOR A PIPE WITH A ROUGH SURFACE

Nikuradse did experiments concerning the pressure drop along the circular pipe with the adhesion of sands creating surface roughness. When the thickness δ of the laminar sublayer is higher than the mean stick out height ϵ of the sands, the pipe friction factor λ has the same value as the smooth pipe. On the other hand, when the thickness δ of the laminar sublayer is thinner than the mean stick out height ϵ of the sands, the pipe friction factor λ has a higher value than that of the smooth pipe due to the additive pressure drop by the creation of eddies behind the sands, as shown in Figure 4. In the latter case, roughly speaking, the friction factor λ can be written

$$\lambda = \frac{1}{\left\{1.74 - 2\cdot\log_{10}\left(\frac{2\cdot\epsilon}{D}\right)\right\}^2} \qquad (20)$$

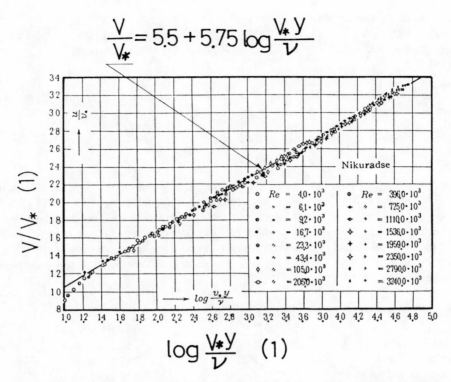

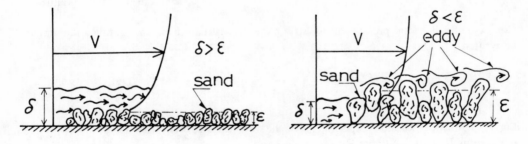

FIGURE 3.   Nikuradse experiments of the velocity profile in the smooth pipe for the high Reynolds number.

FIGURE 4.   Relationship between the laminar sublayer and the height of the sands.

for Re $\geq$ 900/(ϵ/D). The Nikuradse experimental results of the pipe friction factor[3,4] are shown in Figure 5. From the engineering point of view, the practical curves of the pipe friction factor which depends on the relative roughness (ϵ/D) are shown in Figure 6. Detailed information concerning the friction factors for large conduits flowing full is presented in the engineering monographs of Bradley and Thompson (1951).[5]

IV. PIPE FRICTION FACTOR FOR SPECIAL CROSS-SECTIONAL FORMS

A. General Representation of the Pipe Friction Factor

In the case of special cross-sectional forms apart from the circular form, denoting that A(m²) is the cross-sectional area, S(m) is the length of the periphery, Δp (Pa) is the pressure drop for the pipe length L(m), $\overline{V}$(m/s) is the mean fluid velocity in the pipe, and τ_m (Pa) is

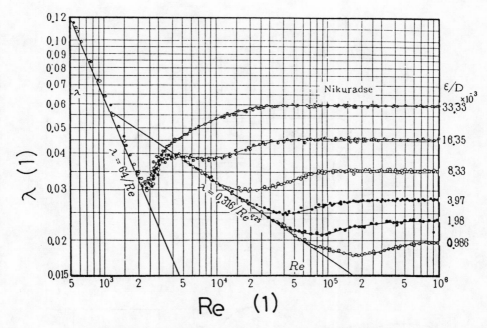

FIGURE 5.   Nikuradse experimental results of the pipe friction factor.

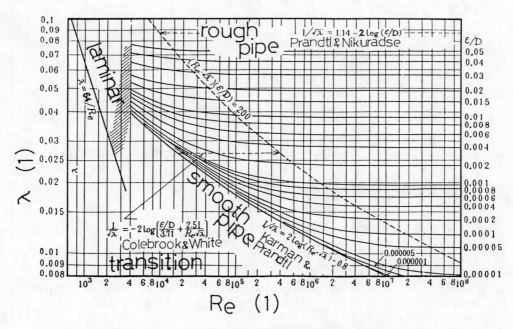

FIGURE 6.   Pipe friction factor depending upon the relative roughness.

the shear stress on the surface of a pipe as shown in Figure 7, we can obtain the following equation

$$A \cdot \Delta p = S \cdot L \cdot \tau_m = S \cdot L \cdot f \cdot \frac{\rho \bar{V}^2}{2} \qquad (21)$$

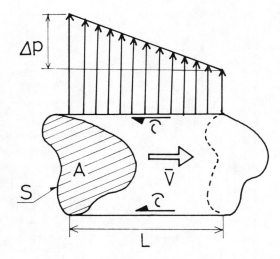

FIGURE 7.   Shear stress on the surface of a pipe.

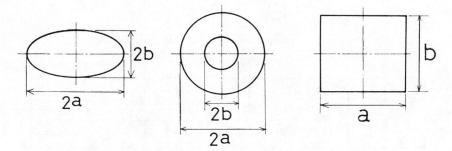

FIGURE 8.   Shapes of the pipes.

Here substituting $f = \lambda/4$ into Equation 21, we can obtain the following equation

$$\Delta p = \frac{f \cdot L}{A/S} \cdot \frac{\rho \overline{V}^2}{2} = f \cdot \frac{L}{m} \cdot \frac{\rho \overline{V}^2}{2} = \frac{\lambda}{4} \cdot \frac{L}{m} \cdot \frac{\rho \overline{V}^2}{2} \qquad (22)$$

and we also define Reynolds number Rem as

$$\text{Rem} = \frac{4 \cdot m \cdot \overline{V}}{v} \qquad (23)$$

where $m(m) = A/S$ is called the hydraulic mean depth. In the case of the circular pipe of diameter D, the hydraulic mean depth is equal to $m = (\pi/4) \cdot D^2/\pi D = D/4$.

B. In the Case of Laminar Flow

The fundamental differential equation of laminar viscous flow in conduits becomes the same form as that of stress distribution for twisting rods of special forms. Then we can apply solutions of stress analysis of elasticity to the problems of fluid flow in the conduits apart from the circular pipe.[6,7]

1. Elliptical Pipe

As shown in Figure 8, in the case of elliptical cross-sectional pipe, the following results were obtained:

velocity distribution

$$V = \frac{\Delta\rho}{2 \cdot \eta \cdot L} \cdot \frac{a^2 \cdot b^2}{a^2 + b^2} \left\{ 1 - \left(\frac{X}{a}\right)^2 - \left(\frac{Y}{b}\right)^2 \right\} \qquad (24)$$

mean velocity

$$\overline{V} = \frac{\Delta p}{4 \cdot \eta \cdot L} \cdot \frac{a^2 \cdot b^2}{a^2 + b^2} = \frac{1}{2} \cdot Vo \; (Vo; \text{center velocity}) \qquad (25)$$

flow rate

$$Qo = \frac{\pi \cdot \Delta p}{4 \cdot \eta \cdot L} \cdot \frac{a^3 \cdot b^3}{a^2 + b^2} = \pi \cdot a \cdot b \cdot \overline{V} \qquad (26)$$

friction factor

$$f = \frac{64}{Rem} \cdot \frac{2 \cdot (a^2 + b^2)}{K^2} \; ,$$

$$K = (a + b) \cdot \left\{ 1 + \frac{1}{4} \cdot \left(\frac{a - b}{a + b}\right)^2 + \frac{1}{64} \cdot \left(\frac{a - b}{a + b}\right)^4 + \ldots \ldots \right\} \qquad (27)$$

2. Coaxial Cylindrical Pipes

For the coaxial cylindrical pipes, we can obtain the following results:

velocity distribution

$$V = \frac{\Delta p}{4 \cdot \eta \cdot L} \left\{ b^2 - r^2 + \frac{a^2 - b^2}{\ln(a/b)} \ln \frac{r}{b} \right\} \qquad (28)$$

radius r_o of the maximum axial velocity

$$r_O = \sqrt{\frac{1}{2} (a^2 - b^2) \Big/ \ln\left(\frac{a}{b}\right)} \qquad (29)$$

mean velocity

$$\overline{V} = \frac{\Delta p}{8 \cdot \eta \cdot L} \left\{ (a^2 + b^2) - \frac{(a^2 - b^2)}{\ln(a/b)} \right\} \qquad (30)$$

flow rate

$$Qo = \frac{\pi \cdot \Delta p}{8 \cdot \eta \cdot L} \left\{ (a^4 - b^4) - \frac{(a^2 - b^2)^2}{\ln(a/b)} \right\} \qquad (31)$$

friction factor

$$f = \frac{64}{Rem} \cdot \frac{(a - b)^2}{(a^2 + b^2) - (a^2 - b^2)/\ln(a/b)} \qquad (32)$$

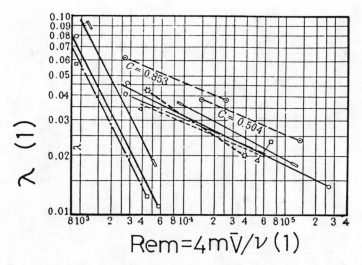

FIGURE 9.   Pipe friction factor for the square, rectangle, triangle, and star pipes.

3. Rectangle Pipe

For the rectangle pipe, we can obtain the following results:

velocity distribution

$$V = \frac{\Delta p}{\eta \cdot L} \left[\frac{1}{2} y^2 + \frac{16b^2}{\pi^2} \left\{ \cos \frac{\pi y}{2b} \cdot \frac{\cosh(\pi x/2b)}{\cosh(\pi a/2b)} - \right. \right.$$

$$\left. \left. - \frac{1}{3^3} \cdot \cos \frac{3\pi y}{2b} \cdot \frac{\cosh(3\pi x/2b)}{\cosh(3\pi a/2b)} + ... \right\} \right] \tag{33}$$

mean velocity

$$\bar{V} = \frac{\Delta p \cdot b^2 \cdot K}{4 \cdot \eta \cdot L} \tag{34}$$

where the values of K depend on the aspect ratio and describe below

a/b	1	2	3	4	5	10	12	100	∞
K	2.253	3.664	4.203	4.498	4.665	5.000	5.059	5.299	5.333

flow rate

$$Qo = \frac{\Delta p \cdot a \cdot b^3 \cdot K}{4 \cdot \eta \cdot L} \tag{35}$$

friction factor

$$f = \frac{64}{Rem} \cdot \frac{8 \cdot a^2}{(a + b)^2 \cdot K} \tag{36}$$

C. In the Case of Turbulent Flow

In this case, the approximate values of λ for circular pipe (symbol $\bigcirc$), coaxial circular pipe (symbol $\circledcirc$), pipe (symbol, $\square$), square/rectangle pipe (symbol $\square$), triangle pipe (symbol, $\triangle$), and the star form pipe (symbol, $\stackrel{\star}{\bowtie}$) are shown in Figure 9. From this figure, the friction

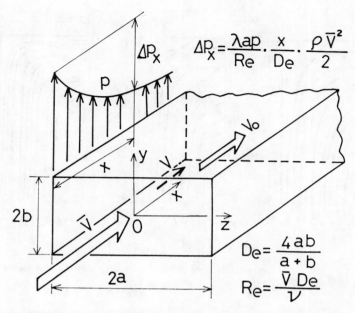

FIGURE 10. Illustration of the entrance length for laminar flow in rectangle duct.

factors $\lambda = 4f$ do not always coincide with the value of λ of the circular pipe due to the complicated flow pattern. Therefore, when the exact value of λ for the special cross-sectional form is needed, one should estimate the value of λ from the model experiments.[8]

V. ENTRANCE LENGTH FOR LAMINAR FLOW IN RECTANGULAR DUCTS

Han (1960)[9] solved the problem of determining the hydrodynamic entrance length in a rectangular channel by the method of linearizing the Navier-Stokes equations for the incompressible laminar flow. The entrance length defined as 99% of the fully developed center-line velocity V_∞ was attained and was calculated for channels of six aspect ratios.

The pressure drops Δp_x (Pa) were also calculated by an equation defined as

$$\Delta p_x = \frac{\lambda ap}{Re} \cdot \frac{X}{De} \cdot \frac{\rho \cdot \bar{V}^2}{2} \qquad (37)$$

where λap (1) is the apparent friction factor, De (m) is an equivalent diameter defined as $De = 4ab/(a + b)$ as shown in Figure 10, $\bar{V}$ (m/s) is the mean axial velocity of fluid, and Re (1) is Reynolds number defined as $Re = \bar{V} \cdot De/\upsilon$.

Figure 11 shows the calculated equi-axial velocity contours $V/\bar{V}$ (1) in a channel of 2:1 aspect ratio for Reynolds numbers, Re = 21.7 (X/De), Re = 64.1 (X/De), and Re = 581 (X/De), respectively. Figure 12 shows a relationship between the axial-velocity ratio Vo/$\bar{V}$ and the ratio (X/De·Re) of the axial distance in a channel. In this figure, the dotted line shows a locus for (Vo/V_∞) = 0.99. Figure 13 shows an apparent friction factor λap (1) of Equation 37 for the entrance lengths X in rectangular ducts of various aspect ratios (a/b).

VI. FLOW PAST A CYLINDER AND A SPHERE

A. Flow Past a Cylinder (Blasius Series)

Figure 14 shows the velocity distribution in the boundary layer on a circular cylinder of radius R. As is well known, the ideal velocity distribution in inviscid irrotational flow past

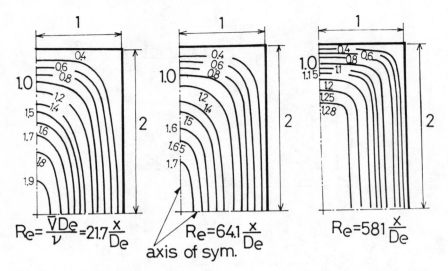

FIGURE 11.   Calculated equi-axial velocity contours in a rectangular channel for the various Reynolds numbers.

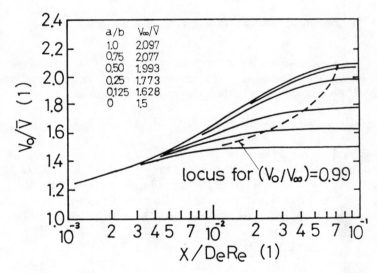

FIGURE 12.   Relationship between the axial velocity ratio and the axial distance in the channel.

a circular cylinder of radius R and free-stream velocity $\overline{V}o$ parallel to the x-axis can be written

$$\overline{V}_{\infty}(\theta) = 2 \cdot \overline{V}o \cdot \sin\theta \tag{38}$$

where a symbol θ is the angle measured from the stagnation point.[10] Then from the distribution of the shear stress

$$\tau_o = \eta(d\overline{V}(\theta)/dy)o$$

on the surface of the cylinder, the stagnation point of the separation which is corresponding to $\tau o = 0$ is given as

$$\theta s = 1.90 \text{ rad}$$

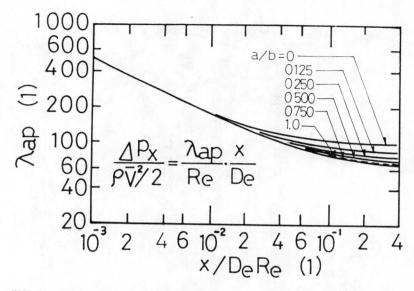

FIGURE 13. Apparent friction factor for the entrance lengths in the rectangular ducts or channels.

B. Viscous Flow Around a Sphere at Low Reynolds Number (Rep $\leq$ 40)

Figure 15 shows the streamlines around a sphere at a low Reynolds number about a sphere, namely, Rep = 5, 10, 20, and 40. The streamlines were calculated numerically by Jenson.[11] These flow patterns which are changed with Reynolds number Rep are very important for the understanding of the separation mechanisms of a fine solid particle in the gas flow.

C. Flow Past an Accelerated Moving Spherical Particle

Hilprecht (1976)[12] made the numerical analysis of unsteady motion of a solid particle in fluid. In order to describe the mathematical analysis of the motion of a particle, he introduced the dimensionless parameters defined as

$$\text{Reynolds number} \quad \text{Rep} = W_\infty \cdot Xp/v \tag{39}$$

$$\text{Reynolds number} \quad \overline{\text{Rep}} = Ws_\infty \cdot Xp/v \tag{40}$$

$$\text{Fourier number} \quad Fo\tau = v \cdot t/Xp^2 \tag{41}$$

$$\text{Galilei number} \quad Ga = g \cdot Xp^3/v^2 \tag{42}$$

where W_∞ is the relative velocity between the particle velocity Up and the fluid velocity Vo and Ws_∞ is the relative velocity between the terminal velocity Wsg of the particle and the fluid velocity Vo. Figure 16 shows the streamlines ψ^*/Rep around an accelerated solid particle of diameter Xp which falls with slow velocity in fluid under the conditions $\overline{\text{Rep}} \leq 0.1$ and $Fo\tau = 10^{-2}$. Hilprecht did the numerical analysis with Navier-Stokes equations for $10^{-6} \leq Ga \leq 10^{-1}$ and for $10^{-3} \leq \rho^* \leq 10^1$, where ρ^* means ρ/ρ_p. Figure 17 shows the streamlines around a solid particle of diameter Xp which falls with slow, steady velocity in fluid under the condition $\overline{\text{Rep}} \leq 0.1$ and $Fo\tau \rightarrow \infty$. From these two figures, it can be seen that the streamlines around an accelerated solid particle for the Fourier number $Fo\tau = 10^{-2}$ become parallel lines at the distance Xp from the surface of the particle, while the streamlines for steady motion become the parallel lines at the distance $2 \cdot Xp$ from the surface of the particle. Therefore, the volumetric flow around the particle in case of the accelerated motion of the solid particle near the wall surface is more in comparison with the case of the steady

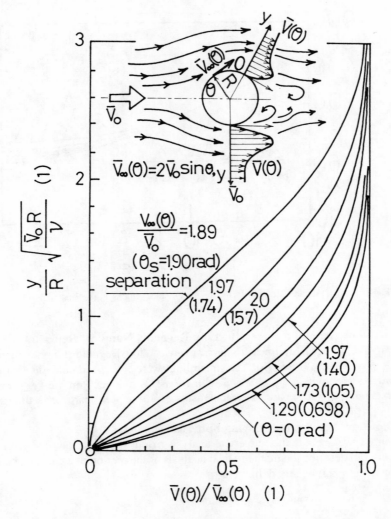

FIGURE 14.   Velocity distributions (Blasius series) in the boundary layer on a circular cylinder.

motion on the same Reynolds number. Figure 18 shows the numerical results of the distributions of the tangential velocity $V_\theta^* = V\theta/W_\infty$ for the radial distance $r^* = r/R$ at $\theta = 90°$. For $Fo\tau = 10^{-4}$ the velocity distribution nearly approaches the distribution of the potential flow and at $r^* = r/R \doteqdot 1$ the value of V_θ^* becomes nearly 1.5. For the falling particle with the very high velocity, the streamlines around an accelerated solid particle are shown in Figures 19, 20, 21, and 22. In this calculation Hilprecht applied the drag coefficient C_D as

$$C_D = \frac{24}{Rep} + \frac{5.48}{Rep^{0.573}} + 0.36 \tag{43}$$

Figure 23 shows the separation angle θs at $Ga = 500$. In this figure other numerical and experimental results are shown, and also in the case of steady motion, the separation angle θs for $20 \leqq \overline{Re}p \leqq 400$ can be written

$$\theta s = \left[\frac{\ln(\overline{Re}p - 3)}{2.3 \times 10^{-4}} \right]^{0.45} \tag{44}$$

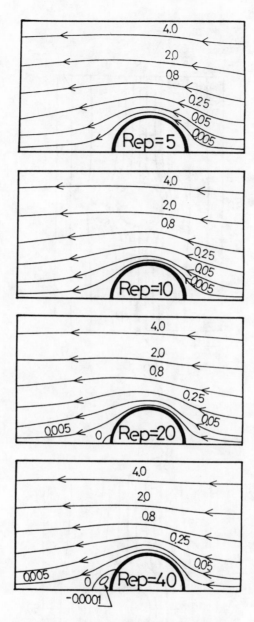

FIGURE 15.   Stream lines around a sphere at the
low Reynolds number.

Figure 24 shows the distance L of the wake behind the particle at Ga = 500. The symbols
used in this figure are the same as those of Figure 23. A dotted line for the steady state for
$20 \leq \overline{Re}p \leq 130$ can be written

$$\frac{L}{X_p} = 0.88 \cdot \ln(\overline{Re}p + 22) - 3.29 \tag{45}$$

VII. KÁRMÁN VORTEX STREET

A. Phenomena of Kármán Vortex Street

As is known, a regular Kármán vortex street is observed only in the region of Reynolds

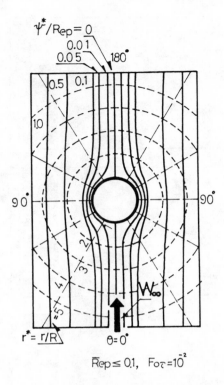

FIGURE 16.   Stream lines around the accelerated particle falling with slow velocity in fluid.

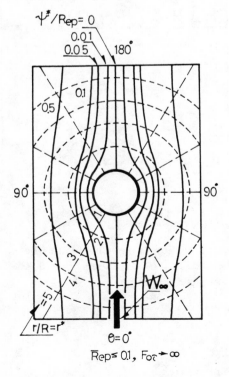

FIGURE 17.   Stream lines around the slow steady velocity in fluid.

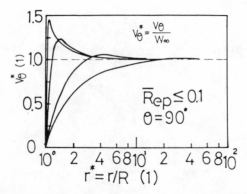

FIGURE 18.   Numerical results of the tangential velocity for the radial distance.

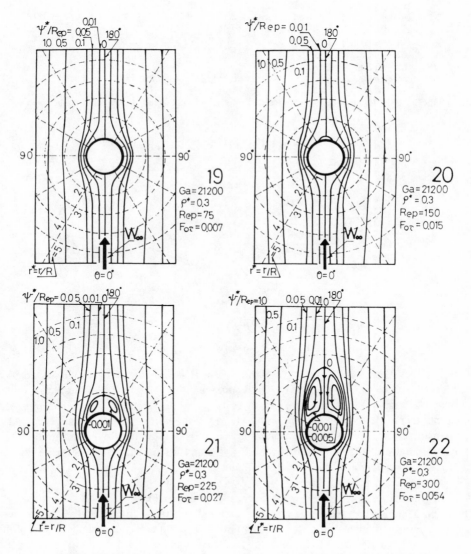

FIGURES 19—22.   Stream lines around an accelerated solid particle with high velocity.

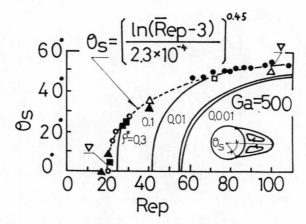

○   Pruppacher,H.R., Le Clair,B.P., Hamielec,A.E.,(1970)
△   Hamielec,A.E., Hoffmann,T.W., Ross,L.L.,(1967)
□   Lin,C.L., Lee,S.C.,(1973)
▼   Rhodes,J.M.,(1967)
▲   Jenson,V.G.,(1959)
■   numerical calculation
●   Taneda,S.,(1959), experimental results

FIGURE 23.   Relationship between the separation angle and the particle Reynolds number.

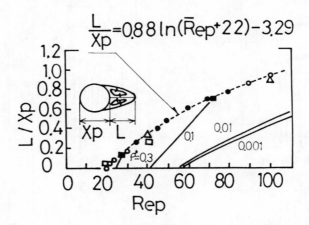

FIGURE 24.   Distance of the wake behind the particle.

number Re = V·D/v from about 60 to 5000, where D is a diameter of a cylinder and Vo is the velocity of gas (air). At lower Reynolds numbers the wake behind a circular cylinder is laminar. At higher Reynolds numbers there is complete turbulent mixing.[13] Here, the Strouhal number is defined as S = f D/Vo, where f is the vortex shedding frequency. Also this dimensionless frequency (Strouhal number) depends only on the Reynolds number. The relationship between Re and S and between C_D, 1/S and Re are shown in Figure 25, where C_D is the drag coefficient of cylinder of diameter D.

B. Vortex Shedding from Circular Cylinders in Sheared Flow

Chen and Mangione[14] investigated the vortex shedding from circular cylinders in shear flow for Reynolds number range Re = V·D/v = 200 to 800. The experiments were done in a suction wind tunnel whose test section was 279.4 mm × 304.8 mm wide. The center line velocity Vc was 3.35 to 8.53 m/s. Then in order to obtain tunnel shear flow, 36 (3.175

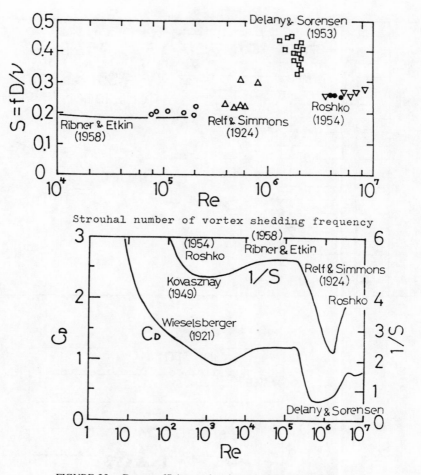

FIGURE 25.   Drag coefficient and reciprocal of Strouhal number.

mm diameter) stainless steel rods were set with appropriate spacing. The velocity profile was written

$$\frac{V}{Vc} = 1 + 0.57\left(\frac{Y}{H}\right) \qquad (46)$$

The turbulence level in shear flow was 4%. The diameters of the test steel cylinder were D = 1.27 or 1.65 mm, respectively. As shown in Figure 26, a relationship between Strouhal number S = f·D/V and Reynolds number Re = V·D/υ could be described for both uniform and shear flow by the following equation

$$S = \frac{f \cdot D}{V} = 0.212 - \frac{4.5}{Re} \qquad (47)$$

where f is the shedding frequency, D is the diameter of the test cylinder, and υ is the kinematic viscosity.

Figure 27 shows an interesting example of the velocity distribution for the shear flow by means of the steel wires by Galanis and Barrows (1966).

C. Visualization of Accelerated Flow Over a Circular Cylinder

Golovkin, Kalyavkin, and Kolkov[15] did experiments concerning the development of eddies behind a cylinder of diameter D on the accelerated flow in the Reynolds number for steady flow as $700 \leq Re_p \leq 9000$.

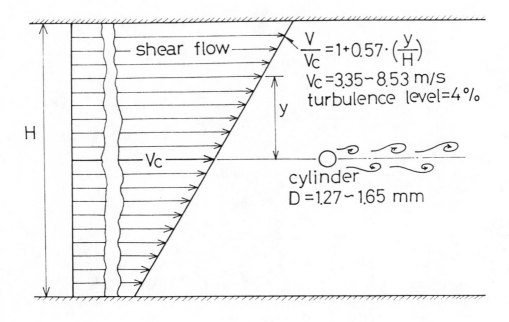

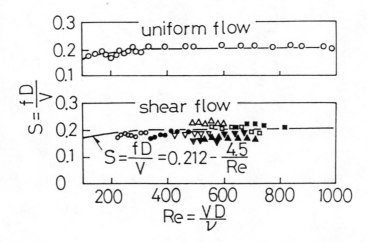

FIGURE 26.   Shear flow and the correlation of Strouhal number with Reynolds number for the both uniform and shear flow.

The flow in the hydrodynamic tube was accelerated from zero to a given steady value Vo of the fluid velocity in accordance with a nearly linear law in a time t corresponding to $\tau = \tau_1 = 1.5$, where τ is defined as $\tau = t \cdot Vo/D$.

In order to detect the flow processes over a cylinder, the Russian authors used the optical visualization method which was connected with a movie camera, a video magnetophone, or a still camera.

Then, using movie stills, they established how the coils of the vortices grow in time t. Let φ be the angle which characterizes the length of a spiral observed in the frames as shown in Figure 28. The number of spiral coils is defined as

$$n = \frac{\varphi}{2 \cdot \pi}$$

(48)

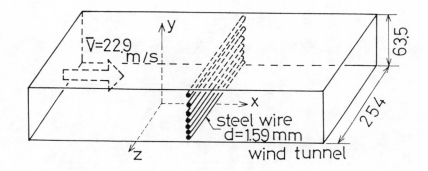

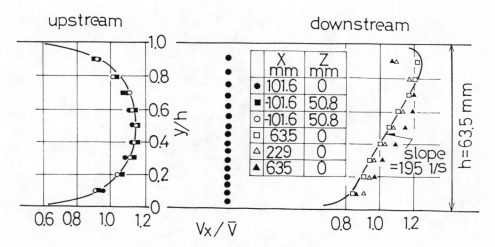

FIGURE 27.   Measured velocity distributions upstream and downstream from the grid.

If we choose the time origin as a representative time corresponding to $\varphi = \varphi o = \pi$, the increase in time of the number of coils of the spiral or its length related by $\Delta\varphi = (\varphi - \varphi o)/2\cdot\pi$ is characterized by a linear dependence as shown in Figure 28. Thus the frequency of formation of spiral coils does not depend on the time for a fixed value of flow velocity, i.e., the Strouhal number defined as

$$Sh = \frac{d\varphi}{dt}\cdot\frac{D}{Vo} = \frac{d\varphi}{d\tau} = \text{const.} \tag{49}$$

The drawings reproduced from photographs show the formation process of the eddy flow in the delta-shaped region. On the surface of the cylinder one can note the formation of a dark region which merges with the image of the model, its appearance characterizing the start of the second flow separation (τo). At $\tau = \tau o + 0.2$, the secondary separation is more clearly manifested and at $\tau = \tau o + 0.49$, it is the form of a vortex with the opposite direction of rotation to that of the accelerated vortex. As the secondary vortex develops, it interacts with the main flow, namely, with the original vortex layer shed by the cylinder surface and forming the accelerated vortex. This is reflected in an increasing local curvature of the primary layer which then leads to the formation of a new (third) flow. This subsequently forms a third vortex at $\tau = \tau o + 0.79$ and $\tau = \tau o + 1.18$ with the same rotation direction as the first one. In this state of the flow development, the second and the third vortices are the main, largest vortex formations in the delta-shaped region occupying approximately equal parts. During the development, the shape of the spirals forming these vortices changes from the ordinary circular form to almost triangular shape ($\tau = \tau o + 1.18$) due to their interaction

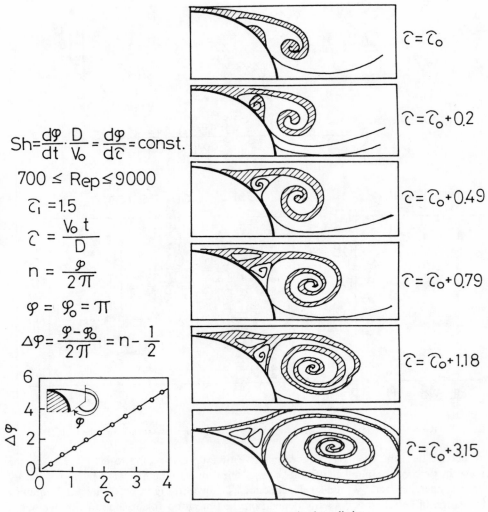

The equations shown in the figure:

$$Sh = \frac{d\varphi}{dt} \cdot \frac{D}{V_0} = \frac{d\varphi}{d\tau} = const.$$

$$700 \leq Re_p \leq 9000$$

$$\tau_i = 1.5$$

$$\tau = \frac{V_0 \, t}{D}$$

$$n = \frac{\varphi}{2\pi}$$

$$\varphi = \varphi_0 = \pi$$

$$\Delta\varphi = \frac{\varphi - \varphi_0}{2\pi} = n - \frac{1}{2}$$

Flow panel labels:

$\tau = \tau_0$

$\tau = \tau_0 + 0.2$

$\tau = \tau_0 + 0.49$

$\tau = \tau_0 + 0.79$

$\tau = \tau_0 + 1.18$

$\tau = \tau_0 + 3.15$

FIGURE 28.   Accelerated flow over a circular cylinder.

with one another and the exterior flow. Then at the time preceding the loss of flow symmetry, the relief of the interval structure of the delta-shaped region is lost, and its structure becomes more homogeneous ($\tau = \tau_0 + 3.15$).

D. Oscillation of the Wake Behind a Flat Plate Parallel to the Flow

Taneda[16] investigated the characteristics of the oscillations of the wakes behind a flat plate parallel to the flow. The lengths (L, mm) and the thickness (H, mm) of the plates are as follows: L = 400, H = 1.0; L = 223, H = 1.0; L = 140, H = 1.0; L = 101, H = 0.50; L = 37, H = 0.24; L = 20, H = 0.16; respectively. Then, in order to create wake oscillations, the plate was pulled by motor and gear systems in the water tank as shown in Figure 29. For the sufficiently small Reynolds number $Re_L = U \cdot L/\upsilon$, the corresponding wake was perfectly laminar and vortex filament was not seen. When Re_L became about 700, the wake began to oscillate sinusoidally some distance downstream. When increasing Reynolds number Re_L, the oscillation of the wake became more and more noticeable and also discrete vortex filaments appeared in the loop of the sinusoidal wave.

Frequency f of the oscillation was measured by the hot-wire anemometer and the Strouhal number f·L/U was found to be nearly proportional to $\sqrt{Re_L}$.

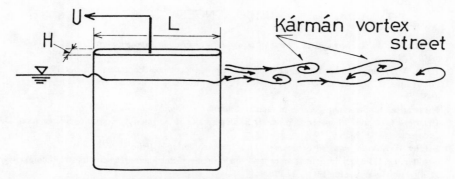

FIGURE 29. Method for the oscillations of the wakes behind a flat plate.

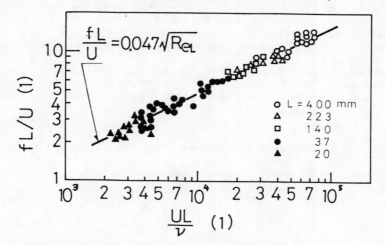

FIGURE 30. Relationship between Strouhal number and Reynolds number.

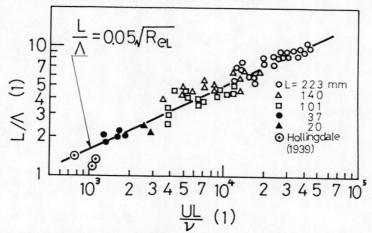

FIGURE 31. Relationship between wave length and Reynolds number.

On the other hand, the wavelength Λ was determined from the photographs taken by the aluminum powder method and a relationship between L/Λ and ReL was obtained as

$$\frac{L}{\Lambda} = \text{const}\cdot\left(\frac{U\cdot L}{v}\right)^{1/2} = \text{const}\cdot\sqrt{ReL}.$$

The experimental results are shown in Figures 30 and 31, respectively.

REFERENCES

1. **Donkin, C. T. B.**, *Elementary Practical Hydraulics of Flow in Pipes*, Oxford University Press, Oxford, 1959.
2. **Itaya, M.**, *Hydraulics* (in Japanese), Japan Society of Mechanical Engineers, Tokyo, 1960.
3. **Nikuradse, J.**, Gesetzmässigkeiten der turbulenten Strömung in Glatten Rohren, *Forschungsheft*, 356, 1932.
4. **Nikuradse, J.**, Strömungsgesetze in rauhen Rohren, *Forschungsheft*, 361, 1933.
5. **Bradley, J. N. and Thompson, L. R.**, U.S. Department of the Interior, Bureau of Reclamation, Denver, Colorado, No. 7, March, 1951.
6. **Müller, W.**, *Mathematische Strömungslehre*, Springer-Verlag, Berlin, 1928.
7. **Schiller, L.**, Über den Strömungswiderstand von Rohren verschieddnen Querschnitts und Rauhigkeitsgrades, *Z. Angew. Math. Mech.*, 13(1), 2, 1923.
8. **Prandtl, L.**, *Essentials of Fluid Dynamics*, Blackie & Son, Ltd., 1963.
9. **Han, L. S.**, Hydrodynamic entrance lengths for incompressible laminar flow in rectangular ducts, *J. Appl. Mech. Trans. Am. Soc. Mech. Eng.*, September, 403, 1960.
10. **Schlichting, H.**, *Boundary Layer Theory*, McGraw-Hill, New York, 1979.
11. **Jenson, V. G.**, Viscous flow around a sphere at low Reynolds number, *Proc. R. Soc. Lond. Ser. A*, 249, 346, 1959.
12. **Hilprecht, L.**, Instationärer Impuls-und Stoffaustausch bei beschleunigten Bewegung von Einzelpartikeln, *VDI-Forschungsh.*, 577, 1971.
13. **Roshko, A.**, Experiments on the flow past a circular cylinder at very high Reynolds number, *J. Fluid Mech.*, 10(3), 345, 1961.
14. **Chen, C. F. and Mangione, B. J.**, Vortex shedding from circular cylinders in sheared flow, *AIAA J.*, 7, 1211, 1969.
15. **Golovkin, V. A., Kalyavkin, V. M., and Kolkov, V. G.**, Optical visualization of accelerated cylinder, *Fluid Dyn. USSR*, 16(2), 266, 1981.
16. **Taneda, S.**, Oscillation of the wake behind a flat plate parallel to the flow, *J. Phys. Soc. Jpn.*, 13, 418, 1958.

Chapter 3

CHARACTERISTICS OF DRAG FORCES ON SOLID PARTICLES

I. INTRODUCTION

Before describing the motion of solid particles and the separation process of the solid particles in a gas flow and in the separation chambers, the author wishes to explain the fundamental characteristics of the drag force D on the solid particles moving in a gas flow.

Generally speaking, the drag force D (N) on the spherical solid particle moving with the velocity Up in the velocity Vo of the gas flow can be written

$$D = C_D \cdot \frac{\pi \cdot X_p^2}{4} \cdot \frac{\rho (V_o - U_p)^2}{2} = C_D \cdot \frac{\pi \cdot X_p^2}{4} \cdot \frac{\rho \, U_r^2}{2} \qquad (1)$$

where a symbol C_D (1) is the drag coefficient, Xp (m) is the diameter of the solid particle, ρ (kg/m^3) is the density of gas, and Ur (m/s) is the relative velocity between the fluid velocity, Vo, and the solid particle velocity, Up, as shown in Figure 1. The drag coefficient C_D of a spherical solid particle of diameter Xp is related to the Reynolds number Rex about a solid particle defined as

$$Rex = \frac{X_p \cdot (V_o - U_p)}{\upsilon} \qquad (2)$$

where υ (m^2/s) is the kinematic viscosity of gas. One example of C_D for a spherical particle is shown in Figure 1. When the form of the solid particle is not a spherical shape, but an irregular shape, we can employ an equivalent diameter as the representative length. When the Reynolds number Rex around the solid particle is less than 4, from the engineering point of view, the drag coefficient C_D becomes theoretically

$$C_D = \frac{24}{Rex} \qquad (3)$$

and drag force (Stokes drag force) D becomes

$$D = 3 \cdot \pi \cdot \eta \cdot X_p \cdot U_r \qquad (4)$$

where η(Pa·s) is the viscosity of gas. In addition to this, we can divide the flow region around the solid particle as a function of Reynolds number

    Stokes flow Rex $\leqq 4$
    Allen flow $4 \leqq Rex \leqq 600$
    Newton flow Rex $\leqq 600$.

The physical flow illustration around the solid particle will be described in detail in a later section.

II. TERMINAL (SEDIMENTATION) VELOCITY OF THE SOLID PARTICLE

Assuming that the solid particle of diameter Xp is falling with velocity Up in a quiet gas, the gravity force G on the solid particle in gas reaches a state of mechanical equilibrium against the drag force D. In this instance, the solid particle is falling with a constant velocity

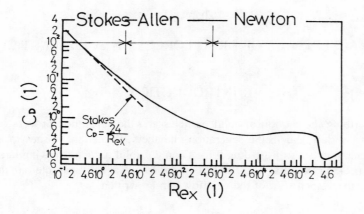

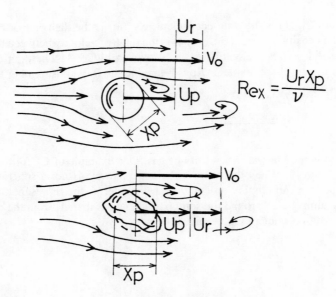

FIGURE 1.   Drag coefficient of a spherical particle and flow illustration around a particle.

Wsg which is called the terminal velocity or sedimentation velocity. This state of mechanical equilibrium can be written

$$G = \frac{\pi \cdot X_p^3}{6} \cdot \rho_p \cdot g = C_D \cdot \frac{\pi \cdot X_p^2}{4} \cdot \frac{\rho \, W_{sg}^2}{2} \tag{5}$$

where ρ_p (kg/m³) is the particle density and Wsg (m/s) is the terminal velocity. Therefore, the equation of the terminal velocity can be written from Equation 5 as

$$W_{sg} = \sqrt{\frac{4 \cdot \rho_p \cdot g \cdot X_p}{3 \cdot \rho \cdot C_D}} \tag{6}$$

Then if the Reynolds number Rex is less than 4, using Equation 3, the terminal velocity Wsg becomes

$$W_{sg} = \frac{\rho_p \cdot g \cdot X_p^2}{18 \cdot \eta} \tag{7}$$

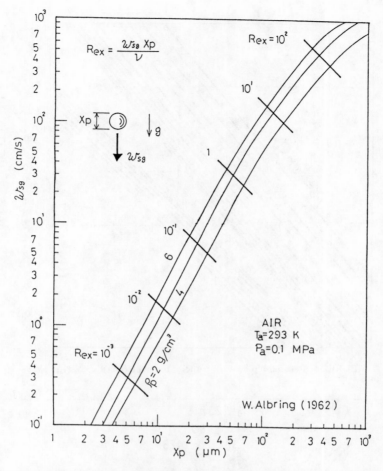

FIGURE 2.   Diagram of terminal velocity of a solid spherical particle with density of a solid particle in quiet air.

Figure 2 is a diagram of a terminal velocity Wsg of a solid spherical particle with density ρ_p in a quiet air with temperature Ta = 293 K and pressure Pa = 0.1 MPa. In this diagram, Reynolds number Rex = Wsg·Xp/υ about a solid particle of diameter Xp based on the terminal velocity Wsg is shown as a parameter.[1] Figure 3 shows a diagram relating the particle diameter Xp (Xp = 2 to 100 μm), terminal velocity Wsg, the particle density ρ_p (ρ_p = 500 to 10,000 kg/m³), and the gas temperature T (T = 273 to 1273 K) of air.[2]

III. DRAG COEFFICIENT

A. Basina and Maksimov Experiment

Basina and Maksimov (1969) did an experiment concerning the drag coefficient C_D of the spherical particle during heat transfer and combustion.[3] The experimental results are shown in Figure 4. In this figure, Tp (K) means the temperature of the solid particle in a burning and heated state, and Ta (K) is the temperature of gas. In addition to this, other empirical equations of the drag coefficients are indicated. This figure is very important for considering the motion of fine solid particles in the combustion chamber, in the waste gas of the blast furnace, and in the high temperature gas.

B. Ingebo Experiment

Ingebo (1956)[4] investigated the motion of clouds of liquid and solid spheres accelerating

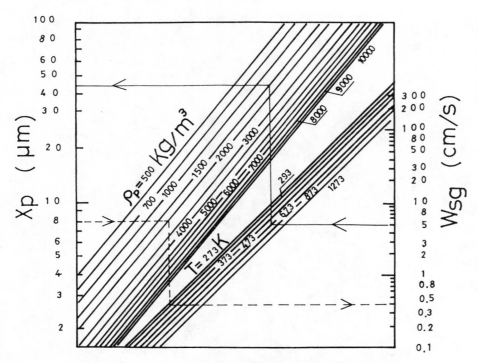

FIGURE 3.   Relationship between particle size and terminal velocity in air.

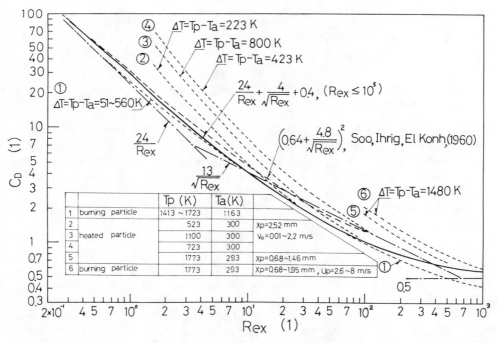

FIGURE 4.   Drag coefficient of the spherical particle during heat transfer and combustion.

in an air stream over a range of air stream pressure (42.9 to 166 kPa), temperature (277 to 644 K), and velocity conditions (30.5 to 54.9 m/s). Diameter Xp and velocity Up data for individual droplets and solid spherical particles in the clouds were obtained with a high

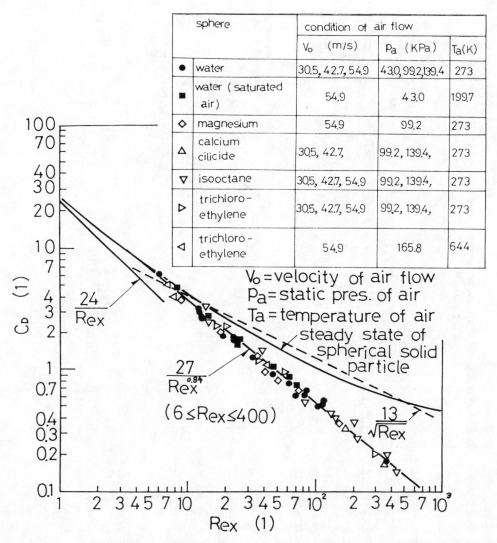

sphere		condition of air flow		
		V_o (m/s)	P_a (KPa)	T_a(K)
●	water	30.5, 42.7, 54.9	43.0, 99.2, 139.4	273
■	water (saturated air)	54.9	43.0	199.7
◇	magnesium	54.9	99.2	273
△	calcium cilicide	30.5, 42.7,	99.2, 139.4,	273
▽	isooctane	30.5, 42.7, 54.9	99.2, 139.4,	273
▷	trichloro-ethylene	30.5, 42.7, 54.9	99.2, 139.4,	273
◁	trichloro-ethylene	54.9	165.8	644

FIGURE 5. Correlation of instantaneous unsteady-state drag coefficient based on linear acceleration with Reynolds number (Ingebo experiment).

speed camera developed at the NACA Lewis Laboratory. From these experimental results, linear accelerations of spherical particles (X_p = 20 to 120 μm) could be determined. Also instantaneous drag coefficients C_D for droplets (isooctane, water, and trichloroethylene) and solid spherical particles (magnesium and calcium silicide) are correlated with the particle Reynolds number Rex, as given by the empirical equation

$$C_D = \frac{27}{Rex^{0.84}}$$

(8)

for $6 \leqq Rex \leqq 400$. As acceleration ratios were low, the unsteady-state drag coefficients agreed with steady-state values of drag coefficients, as shown in Figure 5. Also, Ingebo (1954)[5] investigated the motion of isooctane droplets in turbulent flow. He measured the positions of the droplets by a camera developed at the NACA-TN Lewis Laboratory. From his experimental results, the drag coefficient C_D depended on the particle's Reynolds number Rex and the relative velocity Δu of air with respect to droplets and solid spheres. But these

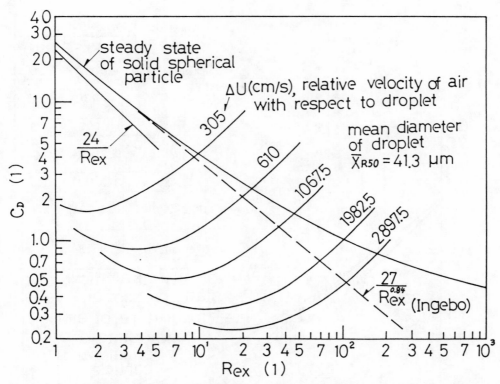

FIGURE 6.   Relationship between the drag coefficient and the Reynolds number for isooctane droplets and solid spheres in a turbulent flow (Ingebo experiment).

values of C_D were lower than that of a solid spherical particle moving with steady motion as shown in Figure 6.

IV. SCHULTZ-GRUNOW AND SAND EXPERIMENT

Schultz-Grunow and Sand (1963)[6] investigated the behavior of solid particles in liquid shear flow near the wall surface. They found that solid particles (Xp = 0.076 to 0.685 mm) accumulated in the shear flow at the region which was represented by a function of a flow Reynolds number about a solid particle. They also measured the lift force Fy on the solid particles (Xp = 0.274 mm) at the various positions in the liquid shear flow for a rectangular pipe (2R = 3 mm) and for a circular pipe (2R = 4 mm). As shown in Figure 7, the lift force Fy on the particles between $0 \leqq y/R \leqq 0.4$ acted toward the center direction and the lift force Fy on the particles between $0.4 \leqq y/R \leqq 1.0$ acted toward the wall direction. In this figure non-dimensional profile of laminar flow

$$V = 1 - \left(\frac{y}{R}\right)^2$$

was shown and Vm was flow velocity at the center of pipe. However, it is noteworthy that this lift force Fy is too small to blow up the solid particles from the wall surface in laminar shear flow of gas, because the gravity force Gs of the solid particle is much heavier than the lift force Fy in gas flow.

V. RELATIONSHIP BETWEEN THE TERMINAL VELOCITY, PARTICLE NUMBER, AND PARTICLE VOLUME CONCENTRATION

From an engineering point of view, we always handle large numbers of particles, a powder,

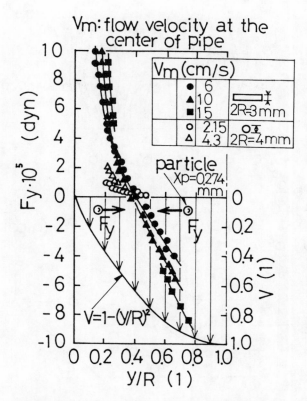

FIGURE 7.   Behavior of solid particles in liquid shear flow (Grunow and Sand).

or solid particles, not just one particle. Therefore, it is very important to consider the behavior of the terminal velocity Wsgm for a group of the particles in gas under the hydrodynamical effects.

As mentioned above, hydrodynamical effect does not affect moving particles in air or in liquid if the distance l between particle and particle, as shown in Figure 8, is longer than about 10Xp, at least. In order to derive a relationship between particle number n per length L and volume concentration, we assume that particles of diameter Xp are arranged at regular intervals, as shown in Figure 8. From this geometrical configuration, we can obtain a relationship

$$L = (n - 1) \cdot 10 \cdot Xp,$$

and the total particle volume Vp per volume L^3 is equal to

$$Vp = \frac{\pi \cdot X_p^3 \cdot n^3}{6} = \frac{\pi \cdot n^3}{6} \cdot \left\{ \frac{L}{10 \cdot (n - 1)} \right\}^3$$

Therefore, volume concentration Cv can be obtained as

$$\frac{\text{particle volume Vp}}{L^3} = \frac{\pi \cdot n^3}{6000 \cdot (n - 1)^3} = \frac{\pi}{6000} \cdot \left(\frac{n}{n - 1} \right)^3$$

From this equation, the particle number n becomes very large, and the maximum volume

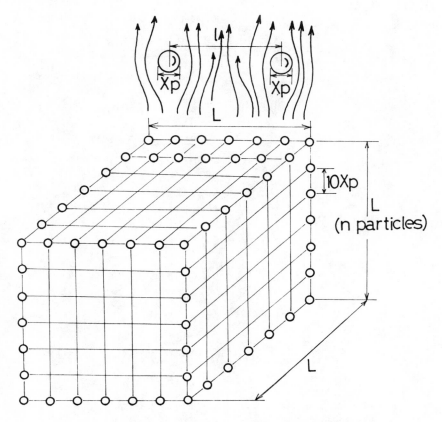

FIGURE 8.   Illustration of particle numbers and volume concentration.

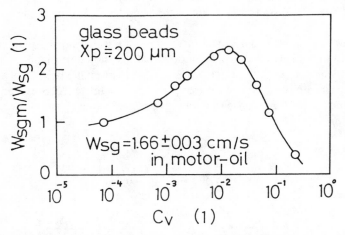

FIGURE 9.   Effect of the concentration of glass beads concerning the terminal velocity in motor oil.

concentration Cv which may not be affected by the hydrodynamical force of moving particles on each other, becomes $Cv \doteqdot \pi/6000 = 5.23 \times 10^{-4}$. Figure 9 shows the effect of the concentration of the glass beads of diameter $Xp = 200$ μm concerning the terminal velocity Wsgm in the motor oil by Johne experiments (1966).[7] The terminal velocity Wsg of one particle in the motor oil was 1.66 ± 0.03 cm/sec. From this figure, one will find that Wsgm/

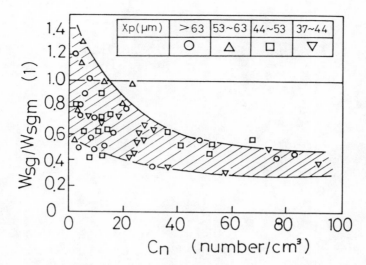

FIGURE 10.   Effect of particle numbers of the fly-ash concerning the terminal velocity in quiet gas.

Wsg $\doteqdot$ 1 corresponds to the volume concentration Cv $\doteqdot$ 10^{-4}. Then increasing the volume concentration of the glass beads to 10^{-2}, the value of Wsgm/Wsg is increased due to the hydrodynamical effect of the particles on each other.

Figure 10 shows the effect of the particle numbers of fly-ash of density ρ_p = 2.1 g/cm³ concerning the terminal velocity Wsgm in quiet gas by Fujita and Ogawa (1980).[8] In spite of the particle diameter Xp in this experiment, the minimum value of Wsg/Wsgm corresponding to number concentration Cn (number/cm³) = 60 to 80 becomes 0.35 to 0.6.

REFERENCES

1. **Albring, W.,** *Angewandte Strömungslehre,* Verlag von Theodor Steinkopff, 1962.
2. **Gordon, G. M., and Peisahov, I. P.,** Pyleulavluvanie i Ochistka Gazov v Tsvetnoj Metallurgii, *Metallurgiia,* 1977.
3. **Basina, I. P. and Maksimov, I. A.,** Issledovanie Aerodinamitseskov Soprotivleniia Steritcheskoj Tchastitsy Pri Teploobmene i Gorenii, *Teploenergetika,* 16(1), 75, 1969.
4. **Ingebo, R. D.,** Drag coefficients for droplets and solid spheres in clouds accelerating in air streams, *NACA Tech. Note,* 3762, 1956.
5. **Ingebo, R. D.,** Vaporization rates and drag coefficients for isooctane sprays in turbulent air streams, *NACA Tech. Note,* 3265, 10, 1954.
6. **Schultz-Grunow, F. and Sand, P.,** Ein Entmischungseffekt in Suspensionen bei wandnaher Scherströmung, *Z. Angen. Math. Mech.,* 43, (1/2), 170, 1963.
7. **Johne, R.,** Einfluß der Konzentration einer monodispersen Suspension auf die Sinkgeschwindigkeit ihrer Teilchen, *Chem. Ing. Tech.,* 38(4), 428, 1966.
8. **Fujita, Y. and Ogawa, A.,** Über den Flussigkeitswiderstand von der Sikenden Partikelchengruppe im Schwerkraftfeld, *J. Coll. Eng. Nihon Univ.,* 21-A, 149, 1980.

Chapter 4

MOTION OF SOLID PARTICLES IN FLUIDS

I. INTRODUCTION

This chapter describes the characteristics of coefficients of drag forces acting on solid particles in a gas flow for solid particles of steady and accelerating motion, and for the motion of the state of burning particles in a high temperature gas. Therefore, the author wants to explain the physical phenomena of the motion of the solid particles in the flowing gas. To understand the characteristics of the drag force on solid particles and on the motion of solid particles in a gas flow, it is very important for the design of the dust collector and for a fundamental investigation of the separation process of the fine solid particles in a separation chamber.[1]

Figure 1 shows the illustration of the flow phenomena around moving solid particles. From this figure, you will find that the motion of solid particles strongly depends on the Reynolds number Rex around the solid particle, namely, Stokes flow (Rex $\leq$4), Allen flow (4 $\leq$ Rex $\leq$ 600), and Newton flow (Rex $\geq$ 1000), respectively.

When the Reynolds number Rex becomes a large value from Allen to Newton flow, wakes (eddy motion of fluid) are created on the surface of the solid particles. Consequently, in addition to the turbulent velocity $\sqrt{\overline{v^2}}$ (m/s) on the mean velocity Vo (m/s) of the gas flow, the wakes strongly interfere with the motion of solid particles. For the physical and mathematical phenomena of the motion of solid particles in the gas flow, please refer to the series papers of Torobin and Gauvin, "Fundamental aspects of solids-gas flow," I — VI. *Can. J. Chem. Eng.*, August 1959 to June 1961; to Soo, *Fluid Dynamics of Multiphase Systems*, Blaisdell Publishing, 1967; to Boothroyd, *Flowing Gas-Solids Suspensions,* Chapman & Hall, 1971; to Richardson, *Aerodynamic Capture of Particles*, Pergamon Press, 1960; to Levich, *Physico-chemical Hydrodynamics,* Prentice-Hall, 1962; or to Davies, *Aerosol Science,* Academic Press, 1966.

II. SOLID PARTICLES IN TURBULENT FLOW

Soo, Ihrig, and Kouh[2] investigated the statistical properties of turbulence conveyance and the diffusion of solid particles (glass beads) of diameter Xp = 100 and 200 μm in the mean gas flow velocity Vo = 6.1 to 30 m/s in a square duct of 76.3 mm. Figure 2 shows the intensity of turbulence $\sqrt{\overline{v^2}}$/Vo (1) and the Lagrangian integral scale L (m), where $\sqrt{\overline{v^2}}$ (m/s) is the mean time fluctuating velocity of the turbulent flow. From this figure, one will find that the intensity of the turbulence $\sqrt{\overline{v^2}}$/Vo is not influenced by solid particle loading within 0.908 to 1.82 kg/min due to enough intervals of particles to each other in a gas flow. Figure 3 shows the intensity (mean fluctuating velocity) $\sqrt{\overline{u_x^2}}$ and $\sqrt{\overline{u_y^2}}$ of solid particles (glass beads) for axial and perpendicular directions for the particle size Xp = 100 to 200 μm, respectively. From this figure, one will recognize that the intensity of solid particles is larger than that of air flow turbulence. This phenomenon is very important for considering the separation mechanisms of fine solid particles in turbulent flow in a separation chamber.[3-5] The behavior of fine solid particles in the turbulent air flow strongly depends on the scale of the turbulence, for example, Kolmogoroff micro scale η^*, the smallest dissipation eddy length λ, or integral scale L.[6,7]

Figure 4 shows the diffusion coefficient Dp (m^2/s) of solid particles and the turbulent eddy diffusivity D (m^2/s) of air flow for Reynolds number Re = 3.5 $\times$ 10^4 to 1.5 $\times$ 10^5 in the duct. Solid particle (glass beads) loading is 0.091 to 0.181 kg/min and the mixture

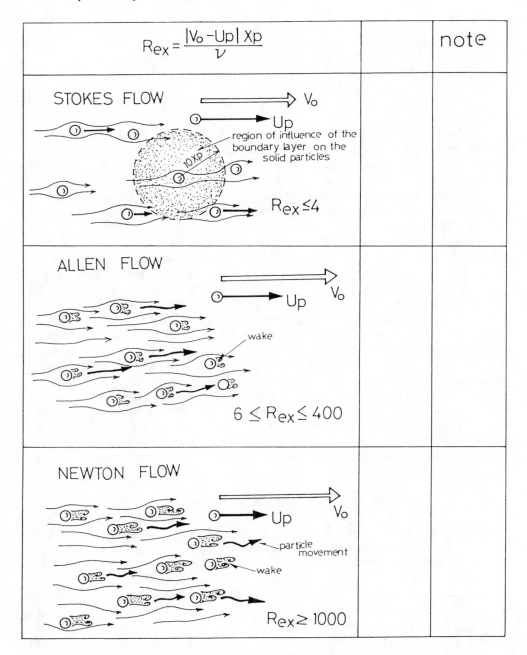

FIGURE 1.   Illustrations of the flow phenomena around moving solid particles.

ratio is 0.01 to 0.06. The value of Dp for the solid particles is larger than D of the turbulent air flow.

Figure 5 shows a dimensionless value of Dp/D for a dimensionless number P defined as

$$P = Fr^{-2} \cdot Rex \cdot \left(\frac{L}{Xp}\right)^2 \cdot Re \cdot 10^{-5} \qquad (1)$$

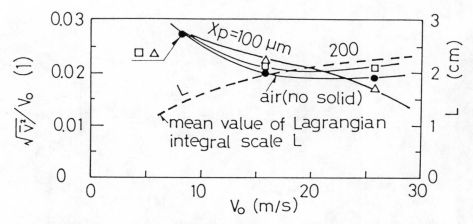

FIGURE 2.   Turbulent intensity of air flow with and without solid particles (glass beads) and scale of turbulence.

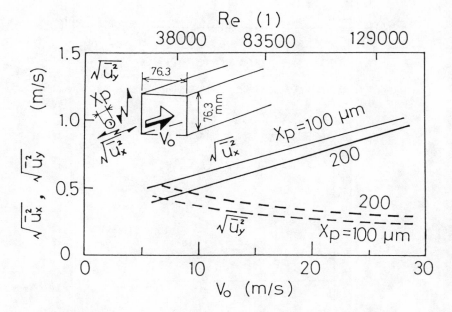

FIGURE 3.   Intensity of the motion of solid particles (glass beads) in ducts.

where Fr and Rex are defined specially as

$$Fr = \frac{\bar{V}^2}{g \cdot Xp}$$

$$Rex = \frac{Xp \cdot \sqrt{\bar{u}^2}}{v}$$

respectively. An empirical equation of Dp/D in duct can be written for the particle diameter Xp = 50 to 200 μm as

$$\frac{Dp}{D} \doteq 0.055 \cdot P + 0.044 \qquad (2)$$

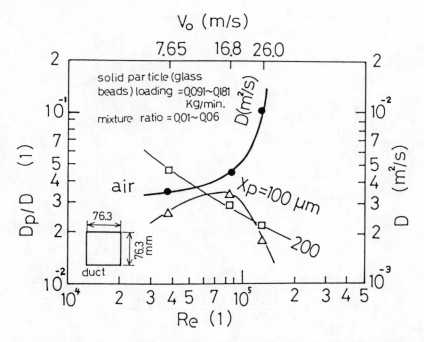

FIGURE 4.   Diffusion coefficient of solid particles and turbulent eddy diffusivity in ducts.

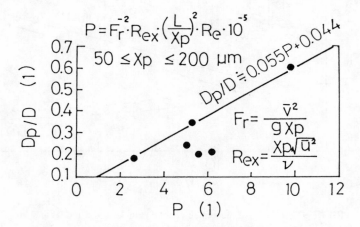

FIGURE 5.   Dimensionless value of Dp/D solid particles in ducts.

Then, Soo et al. numerically calculated the duct moving length X (m) required to accelerate solid particles (glass beads in air at room condition) to near the mean velocity of the air flow.

The mathematical description is as follows. Assuming that the solid particles are introduced at nearly zero axial flow velocity, they are accelerated by the air flow over a length X to nearly the same mean velocity Vo of air flow. The length X is determined by integrating a relation

$$\frac{d(Vo - Up)}{dt} = F \cdot (Vo - Up) \tag{3}$$

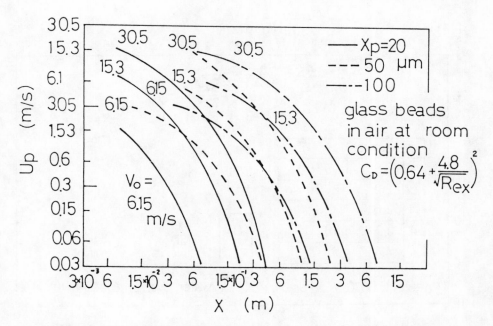

FIGURE 6.   Duct length required to accelerate the solid particles to near the mean velocity of air flow at room condition.

where (Vo − Up) is the velocity difference between the constant mean velocity Vo of gas flow and the mean velocity Up of a solid particle, and also a symbol F is defined as

$$F = \frac{3 \cdot C_D \cdot \rho}{4 \cdot Xp \cdot \rho_p} \cdot (Vo - Up) \tag{4}$$

which is a measure of the relation between viscous drag force and the inertial force of the solid particle and gas flow.

While the moving length dX traversed by the solid particle in time dt is Up dt, eliminating time dt from Equation 3 and followed by integration, the following equation can be obtained as

$$X = \int_0^{Vo-Up} \frac{Up}{F \cdot (Vo - Up)} \cdot dUp \tag{5}$$

Approximately, the drag coefficient C_D is applied defined as

$$\sqrt{C_D} = 0.64 + \frac{4.8}{\sqrt{(Vo - Up) \cdot Xp/\upsilon}} \tag{6}$$

The result of the numerical integration is shown in Figure 6 by Soo et al.

III. LOCI OF WATER DROPLETS IN A RECTANGLE DUCT

Figure 7 shows the motion of a water droplet of the diameter Xp = 360 μm in a horizontal rectangular pipe by Watzel (1970).[8] Each symbol indicates the water droplet's position as

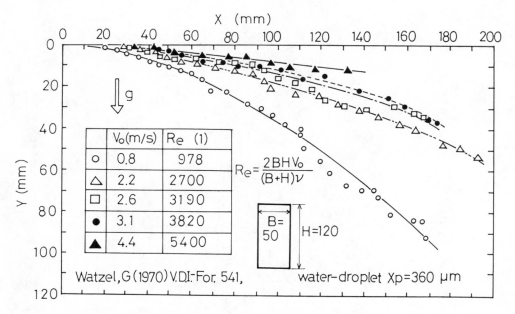

FIGURE 7.   Loci of the water droplets in a rectangular duct.

recorded in photographs. Then in order to estimate the flow Reynolds number Re in the rectangular pipe, the hydraulic mean radius m defined as

$$m = \frac{B \cdot H}{2(B + H)}$$

is applied. Therefore the Reynolds number Re can be written as

$$Re = \frac{4 \cdot m \cdot Vo}{v} = \frac{2 \cdot B \cdot H \cdot Vo}{(B + H) \cdot v} \qquad (7)$$

With an increased Reynolds number, it is very difficult to distinguish the loci of the droplets between Re = 2700 and 5400. When the flow Reynolds number Re is increased from Re = 978 to 5400, the intensity of the turbulent flow is increased nonhomogeneously in the cross-section of the rectangular pipe.[9-11] Consequently the random motion of a water droplet under the force of gravity occurs by the influence of the high-turbulent velocity and the secondary flow near the corner in the rectangular pipe.

Therefore, it is very important to note that the motion of fine solid particles in the high turbulent flow under gravity force may be independent of the flow Reynolds number. Rumpf (1960)[12] noted that the motion of solid particles (glass spheres) of diameters Xp = 157, 97, and 43 μm was not influenced by the effect of the turbulence in air velocity Vo = 282 m/s.

IV. DEPOSITION OF SOLID PARTICLES IN THE TURBULENT GAS FLOW

A. Mechanism of Particle Transport

The physical mechanisms which may contribute to the transport of suspended fine solid particles to the wall of the duct are thermal gradients, gravitational force, diffusional force, electrostatic force, and inertial force.

In the experiments of Friedlander and Johnstone (1957), gravitational effects were eliminated by using vertical tubes. Also the movement of the solid particle by Brownian diffusion

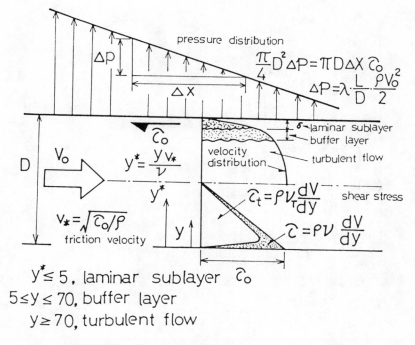

$$\frac{\pi}{4}D^2 \Delta P = \pi D \Delta X \, \tau_o$$

$$\Delta P = \lambda \cdot \frac{L}{D} \cdot \frac{\rho V_o^2}{2}$$

$$y^* = \frac{y \, V_*}{\nu}$$

$$V_* = \sqrt{\tau_o / \rho}$$
friction velocity

$$\tau_t = \rho \nu_r \frac{dV}{dy}$$ shear stress

$$\tau = \rho \nu \frac{dV}{dy}$$

$y^* \leq 5$, laminar sublayer τ_o

$5 \leq y \leq 70$, buffer layer

$y \geq 70$, turbulent flow

FIGURE 8.   Illustration of the turbulent gas flow in a pipe.

was made negligible by using solid particles larger than 0.5 μm in high turbulent flow. Therefore, the most important transport mechanism of fine solid particles is based upon the inertial force effect in the pipe's region of turbulence in the fully turbulent flow.

B. Physical Interpretation of the Deposition

The shear stress of a pipe's fluid flow can be expressed as

$$\tau = \rho(\upsilon + \upsilon_T) \cdot \frac{dV}{dy} \tag{8}$$

where υ (m²/s) is the kinematic viscosity of gas and υ_T (m²/s) is the apparent kinematic eddy viscosity, which is a function of the flow Reynolds number and of the distance y from the wall as shown in Figure 8. Assuming that the eddy diffusivities of mass and momentum are equal, the rate of mass transfer can be written

$$M = (Dm + \upsilon_T) \cdot \frac{dC}{dy} \tag{9}$$

where C is the concentration of the particles and Dm (m²/s) is the molecular diffusivity of gas. In the laminar sublayer the eddy diffusivity may be neglected, and in the buffer layer υ_T can be estimated from the velocity distribution.

Then the fine solid particles and the fluid behave alike in the core region of the turbulent flow, but they move differently near the wall surface. Fine solid particles with large inertial force by the fluctuating velocity of turbulent flow in fully turbulent flow region penetrate the quiescent region and reach to the wall surface. On this fundamental basis Friedlander and Johnstone proposed the following model to explain the deposition of fine solid particles. As shown in Figure 9, eddies of turbulence carrying the fine solid particles diffuse from the turbulent core to within one stopping distance S of the wall surface. This is the effective

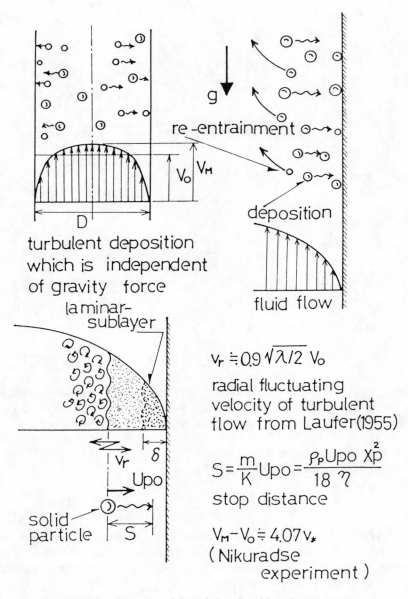

turbulent deposition
which is independent
of gravity force

$$v_r \fallingdotseq 0.9 \sqrt{\lambda/2} \; V_o$$

radial fluctuating
velocity of turbulent
flow from Laufer (1955)

$$S = \frac{m}{K} U_{po} = \frac{\rho_p U_{po} \, X_p^2}{18 \, \eta}$$

stop distance

$$V_M - V_o \fallingdotseq 4.07 v_*$$
(Nikuradse
           experiment)

FIGURE 9.   Illustration of the turbulent deposition of solid particles.

radius of fine solid particles because of their inertial force. It is a distance in which a solid particle with a given initial velocity Upo moves through a stagnant gas and is written by the equation

$$S = \frac{m}{K} \cdot U_{po} = \frac{\rho_p U_{po} \cdot X_p^2}{18 \cdot \eta} \tag{10}$$

Assuming that Upo is equal to the root mean square of the radial fluctuating gas velocity v_r, Friedlander and Johnstone applied the Laufer (1953)[13,14] experimental results of V_r for Reynolds number of 50,000 and 500,000 in 254 mm duct. The radial fluctuating velocity V_r of gas just outside of the buffer layer can be estimated by $V_r \fallingdotseq 0.9 \cdot V_o \cdot \sqrt{\lambda/2}$, where Vo is the mean gas velocity in a pipe and λ is the friction factor. They calculated the stopping

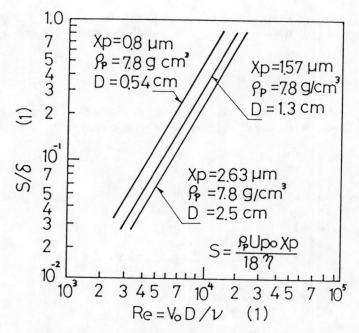

FIGURE 10. Ratio of stop distance to thickness of the laminar sublayer.

distance S for each of the conditions on this basis and this distance S was less than the thickness $\delta \leqq y^* = 5$ of the laminar sublayer. The ratio of the calculated stopping distance S to δ is shown in Figure 10. From this figure, if there were no eddies in the laminar layer, no deposition would have occurred because no solid particles would have diffused within one stopping distance S of the wall surface. Thus it is necessary to assume some fluctuations in the laminar sublayer. Then the rate of transfer can be written

$$M = \epsilon_p \cdot \frac{dC}{dy} \tag{11}$$

but assuming that the eddy diffusivity ϵ_p (m²/s) of the solid particles is equal to v_T, the rate of particle transfer through the laminar sublayer can be written

$$M = v \cdot \left(\frac{y^*}{14.5}\right)^3 \cdot \frac{dC}{dy} \tag{12}$$

where $y^* = y \cdot v_*/v$ is a dimensionless distance. Since the solid particles need to diffuse only to one stopping distance S from the wall, the limits of integration are $C = 0$ at $y^* = S^*$ and $C = Cs$ at $y^* = 5$. If Mo is assumed to be the constant value through the sublayer, then we can obtain the equation

$$Cs = \frac{(14.5)^3 \cdot Mo}{2 \cdot v_*} \cdot \left(\frac{1}{y^{*^2}} - \frac{1}{25}\right) \tag{13}$$

For the buffer layer, the eddy viscosity v_T as given by Lin, Moulton, and Putnam (1953) is

$$\frac{v_T}{v} = \frac{y^*}{5} - 0.959 \tag{14}$$

Substituting Equation 14 into Equation 11 and integrating from C = Cs at y* = 5 to C = Cb at y* = 30 and also assuming that Mo is constant through the buffer layer, so we can obtain the equation

$$Cb - Cs = \frac{24.1 \cdot Mo}{v_*} \qquad (15)$$

While the Reynolds analogy should hold in the turbulent core flow, integrating from C = Cb at y* = 30 to average value C = Ct in the turbulent core, we can obtain the equation

$$Ct - Cb = \frac{Mo}{v_*} \cdot \frac{v}{v_T} \cdot (y^* - 30) \qquad (16)$$

Now adding Equations 13, 15, and 16, we can obtain an equation

Equation 13 + Equation 15 + Equation 16 = Cs + Cb − Cs + Ct −

$$- Cb = Ct = \frac{(14.5)^3 \cdot Mo}{2 \cdot v_*} \cdot \left(\frac{1}{y^{*2}} - \frac{1}{25}\right) + \frac{24.1 \cdot Mo}{v_*} +$$

$$+ \frac{Mo}{v_*} \cdot \frac{v}{v_T} \cdot (y^* - 30) =$$

$$= \frac{Mo}{v_*} \left(\frac{1524}{y^{*2}} - 36.9\right) + \frac{v \cdot Mo}{v_T \cdot v_*} (y^* - 30) \qquad (17)$$

Therefore, a transfer coefficient k of solid particles can be expressed by Equation 18 as follows:

$$k = \frac{Mo}{Ct} = \frac{1}{\frac{1}{v_*}\left(\frac{1524}{y^{*2}} - 36.9\right) + \frac{v}{v_T \cdot v_*} \cdot (y^* - 30)} \qquad (18)$$

Figure 11 shows the deposition of the fine solid particles in 25 mm glass and brass tubes as a function of Reynolds number Re. Figure 12 shows the deposition of the iron particles of diameter Xp = 0.8 μm in a D = 5.8 mm pipe. Increasing Reynolds number Re from 12,600 to 14,900, the transfer coefficient k increases along the pipe axis.

V. MOTION OF THE SOLID PARTICLE IN ALLEN AND NEWTON DRAGS

A. Upward Vertical Motion of the Solid Particle

The upward vertical motion of a solid particle is restrained by the fluid resistance, the collisons with the wall or with other solid particles, and the gravity force of the solid particle. Therefore, the upward vertical motion of the solid particle must be divided into two regions[15]:

1.   When the velocity Upz of the upward vertical motion of a solid particle, which is thrown by an impeller blower, is larger than that of upward fluid velocity Vo, the fluid force Ff on the solid particle brakes.
2.   In case of Upz < Vo, the fluid force Ff pushes up the solid particle.

B. Theoretical Calculation of the Vertical Motion of the Solid Particle Without the Wall Collision

To calculate the motion of a solid particle thrown in the vertical upward direction, the forces acting on the solid particle of mass mp are as follows:

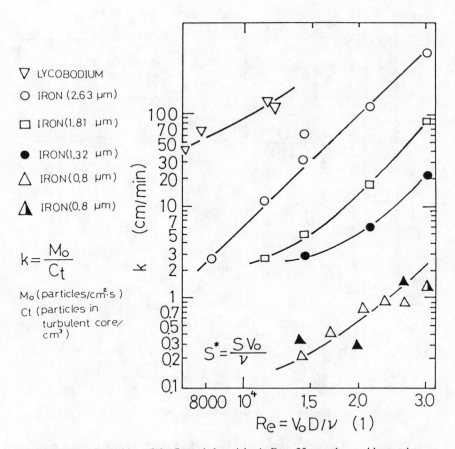

FIGURE 11.   Deposition of the fine soled particles in D = 25 mm glass and brass tubes.

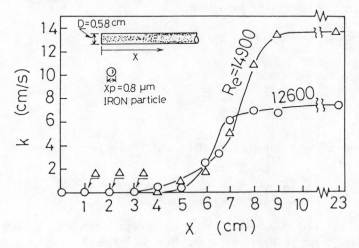

FIGURE 12.   Deposition of 0.8 μm iron particles in a D = 5.8 mm pipe.

1.   The inertia force = mp·dUp/dt
2.   Gravitational force = mp·g
3.   Fluid force = Ff
4.   Buoyancy force

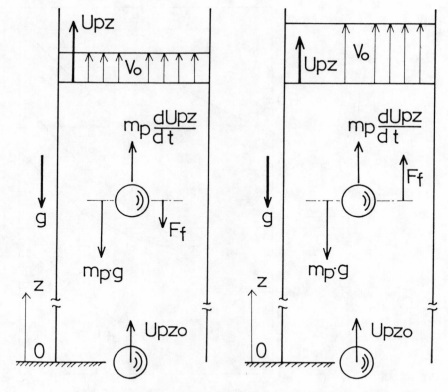

FIGURE 13.   Illustration of the vertical motion of the solid particle in a pipe.

Of those forces, the buoyancy force may be neglected due to the treatment of motion of a solid particle in gas. Then the equation of the upward motion of the solid particle for Upz > Vo showing in Figure 13 can be written

$$mp \cdot \frac{dUpz}{dt} = - Ff - mp \cdot g \qquad (19)$$

that for Upz < Vo can be written

$$mp \cdot \frac{dUpz}{dt} = Ff - mp \cdot g \qquad (20)$$

In those equations, we assume that the fluid force Ff may be written as

$$Ff = C_D \cdot A \cdot \frac{\rho}{2} \cdot U_r^2 \qquad (21)$$

where Ur (m/s) = Vo − Upz, A (m²) is a cross-sectional area and C_D is the drag coefficient which is related to particle Reynolds number Rex = Ur·Xp/υ, as shown in Figure 14. When solid particles are suspended in the vertical upward flow by the state of mechanical equilibrium between the gravitational force and the fluid force, then the equation of motion in this state can be written as

$$mp \cdot g = C_D \cdot A \cdot \frac{\rho}{2} W_s^2 \qquad (22)$$

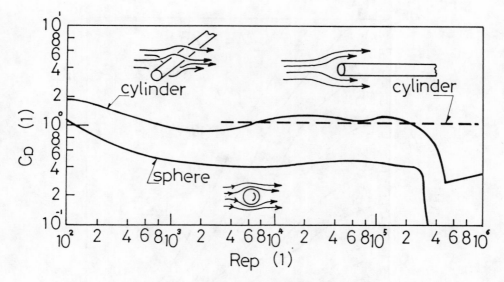

FIGURE 14.   Drag coefficients of a sphere and cylinders.

where the suspension velocity Ws is equal to Ur. Here we can transform the fluid force of Equation 22 with Equation 21 as

$$Ff = mp \cdot g \cdot \frac{U_f^2}{W_s^2} \tag{23}$$

Therefore, Equations 19 and 20 can be transformed as follows:

$$\frac{dU_{pz}}{dt} = -g \left\{ 1 + \left(\frac{Ur}{Ws} \right)^2 \right\} \tag{24}$$

$$\frac{dU_{pz}}{dt} = -g \left\{ 1 - \left(\frac{Ur}{Ws} \right)^2 \right\} \tag{25}$$

Gluth (1971) solved those equations by analog-computer. The numerical results of those calculations for the initial vertical velocities Upzo of the solid particle 20 and 50 m/s for Ws = 15 m/s are shown in Figure 15. In order to compare calculated results with experimental values, Gluth showed examples of experimental results for Ws = 10 and 15 m/s and for Vo = 0 m/s in Figure 16, and for Ws = 7.5 m/s and Vo/Ws = 0.5 in Figure 17. In those figures, the calculated curves coincide with the experimental value for $10^2 \leqq Rep \leqq 10^5$.

VI. FUNDAMENTAL THEORY OF SOLID PARTICLE SEPARATION BY APPLYING ALLEN DRAG LAW

A. Fundamental Respect

The separation of solid particles in a separation chamber[16] depends on the motion of solid particles of various sizes of diameters Xp. For the separation of solid particles, the separation space which correlates the motion of the solid particles in the flow of air is applied. The force acting on solid particles and their movements, depending on the size of the particles, must be controlled in the separation space.

The solid particles corresponding to the cut-size Xc are in a state of mechanical equilibrium and the judgment passing to the fine particles, or to the coarse particles, depends on the

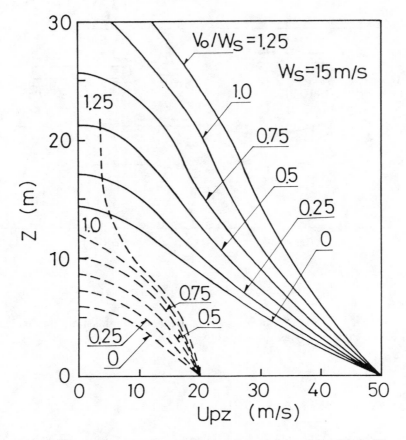

FIGURE 15.   Numerical results of the vertical distance of the solid particle for Ws = 15 m/s.

fluctuating state of the dust-laden gas. The theory of air flow, including solid particles is considered by the separation process of fluid mechanics. The method of calculation of the separation process which was obtained by Sapozhnikov is related to gas velocity with moving solid particles divided into the three separation systems.

B. Separation of the Solid Particle in the Vertical Flow System

As shown in Figure 18, denoting that D(N) is the aerodynamic drag force on the solid particle and G(N) is the gravitational force on the solid particle, we can assume that D is equal to G in the vertical upward gas flow. Then the size of this particle in a state of mechanical equilibrium is generally called cut-size Xc. Therefore, more fine particles than Xc will be moved upward with gas flow and more coarse particles will be separated and move downward from gas flow.

Now it is very difficult to estimate the aerodynamic drag force on solid particles of irregular form. Therefore, we may apply the idea of the equivalent spherical particle which has equal volume and mass. Then the shape factor k is introduced. Sapozhnikov determined the shape factor as follows: k = 1 for a spherical particle, k = 1.1 for an egg-shaped particle, k = 1.5 for a pyramid-shaped particle, k = 1.76 for a slender-shaped particle, and k = 3.8 for a needle-shaped particle.

The equation describing the aerodynamic drag force D acting on a solid particle can be written

$$D = C_D \cdot k \cdot \frac{\pi \cdot X_p^2}{4} \cdot \frac{\rho \cdot (Vo - Up)^2}{2} \qquad (26)$$

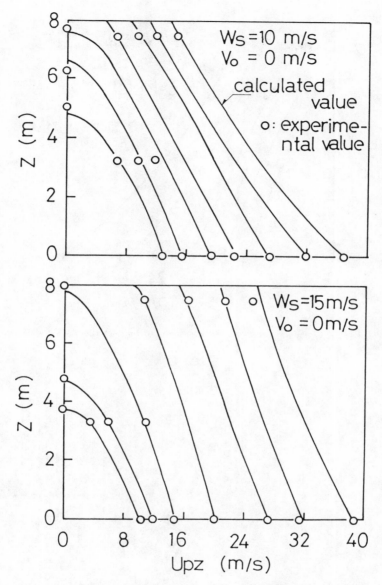

FIGURE 16.   Relationship between calculated and experimental values of z for Vo
= 0 m/s and Ws = 10 and 15 m/s.

where C_D is the drag coefficient, k is the shape factor, Vo is the gas velocity, and Up is the velocity of a solid particle. Here C_D is a function of the Reynolds number Rex = X_p·(Vo − Up)/v about a solid particle. For the gas velocity = 4 to 20 m/s in the separation space and for the particle diameter X_p = 0.1 to 1.0 mm, the Reynolds number is from 50 to 2000. In the above stated flow condition, the drag coefficient can be written as follows:

$$C_D = \frac{13}{\sqrt{Rex}} \qquad (27)$$

where v is the kinematic viscosity of gas. Then the equation of the state of mechanical equilibrium D = G can be written

$$C_D \cdot k \cdot \frac{\pi \cdot X_p^2}{4} \cdot \rho \cdot \frac{|Vo - Up|^2}{2} = \rho_p \cdot g \cdot \frac{\pi \cdot X_p^3}{6} \qquad (28)$$

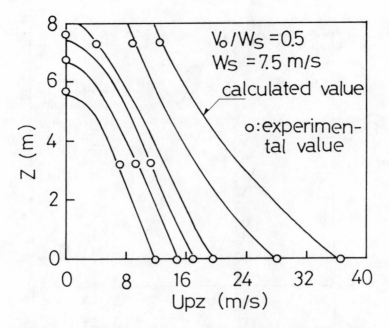

FIGURE 17.   Relationship between calculated and experimental values of Z for Vo/ Ws = 0.5 and Ws = 7.5 m/s.

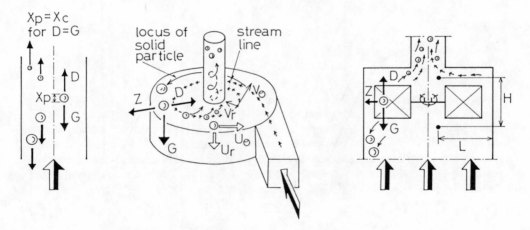

FIGURE 18.   Illustration of the three types of the separators.

where ρ_p is the density of the solid particle. From this mechanical condition, the solid particle is at a standstill, and $|\, Vo - Up\,|$ may be replaced by Vo. We can obtain the equation of the cut-size Xc as

$$Xc = \frac{3}{4} \cdot \frac{k \cdot C_D \cdot \rho \cdot V_o^2}{\rho_p \cdot g} = \left(\frac{39}{4} \cdot \frac{k \cdot \rho \cdot v^{1/2}}{\rho_p \cdot g} \right)^{2/3} \cdot V_o \qquad (29)$$

On the other hand the terminal velocity Wsg corresponding to particle size Xc can be written

$$Wsg = \sqrt{\frac{4}{3} \cdot \frac{\rho_p \cdot g \cdot Xc}{k \cdot C_D \cdot \rho}} \qquad (30)$$

If the solid particle moves in the gas flow, then the velocity Up of the solid particle is equal to

$$Up = Vo - Wsg \tag{31}$$

C. Cyclone Separator

As shown in Figure 18, the solid particle rotates and the gas spirals in the cyclone separator. Then the solid particle moves to the center axis for the condition of the aerodynamic drag force D larger than the centrifugal force Z. On the other hand, the coarser particles move to the outer wall region. From the state of mechanical equilibrium of D = Z, we can obtain the following equation

$$\rho_p \cdot \frac{\pi \cdot X_p^3}{6} \cdot \frac{U\theta^2}{r} = k \cdot C_D \cdot \frac{\pi\, X_p^2}{4} \cdot \frac{\rho \cdot V_r^2}{2} \tag{32}$$

Then the cut-size Xc can be obtained as

$$Xc = \left(\frac{39}{4} \cdot \frac{k \cdot \rho \cdot r \cdot v^{1/2}}{\rho_p \cdot U\theta^2} \right)^{2/3} \cdot Vr \tag{33}$$

where Uθ is the tangential velocity of the solid particle and Vr is the radial velocity of gas. Then assuming that Uθ is equal to Vθ of the gas velocity, Equation 33 becomes

$$Xc = \left(\frac{39}{4} \cdot \frac{k \cdot \rho \cdot v^{1/2} \cdot r}{\rho_p \cdot V\theta^2} \right)^{2/3} \cdot Vr \tag{34}$$

where Vθ is the tangential velocity of the gas.

D. Air Separator

As shown in Figure 18, Z is the centrifugal force on the solid particle, D is the aerodynamic force for upward direction, and G is the gravity force in the separation space.

Solid particles which accept the strong centrifugal force will be thrown to the outer wall surface and will be deposited downward, but the fine particles will move upward with the air flow. Then the cut-size Xc, which can be defined for air separators, is the time required for moving Uz to the vertical distance H, and is equal to that of Ur to the horizontal distance L. Therefore, we can obtain the following equation

$$\frac{H}{Uz} = \frac{L}{Ur} \tag{35}$$

where the axial velocity Uz of the solid particle can be written

$$Uz = Vo - Wsg \tag{36}$$

The radial velocity Ur of the solid particle may be nearly written from the state of mechanical equilibrium between the centrifugal force Z and the aerodynamic drag force D as follows:

$$Ur = \sqrt{\frac{4 \cdot \rho_p \cdot Xc \cdot a_c}{3 \cdot k \cdot \rho \cdot C_D}} \tag{37}$$

where a_c is the mean centripetal acceleration. Then substituting Equation 36 and Equation 37 into Equation 35, we can obtain the equation of the cut-size Xc

$$Xc = \frac{3 \cdot k \cdot \rho \cdot V_0^2}{4 \cdot \rho_p \cdot \left\{ \sqrt{\frac{g}{C_D}} + \frac{H}{L} \cdot \sqrt{\frac{a_c}{C_D}} \right\}^2} \qquad (38)$$

where C_D is a function of the Reynolds number about a solid particle.

Here we calculate the cut-size Xc of Equation 38 for a diameter $D_1 = 5$ m, flow rate Qo $= 100 \times 10^3$ m³/hr, rotational speed n of paddle n $= 180$ RPM, and the mean vertical air velocity Vo $= Qo/(\pi D_1^2/4) = 1.42$ m/s.

If we choose $r = D_1/3$ as a representative radius, so the mean centripetal acceleration $a_c = V_\theta^2/r = 3 \cdot V_\theta^2/D_1 = 3 \cdot D_1 \cdot (\pi \cdot n/90)^2 = 592$ m/s². Now assuming that the density ρ_p of the particle is $\rho_p = 2 \times 10^3$ kg/m³, density ρ of air is $\rho = 1.2$ kg/m³, shape factor is $k = 1$, horizontal length is $L = 2$ m, vertical distance is $H = 1.5$ m, and Reynolds number Rex about a solid particle is Rex $= 100$, so we can obtain the cut-size as follows:

$$Xc = \frac{3 \times 1.2 \cdot (1.42)^2}{4 \times 2 \times 10^3 \left\{ \sqrt{\frac{9.8}{1.3}} + \frac{1.5}{2} \cdot \sqrt{\frac{592}{1.3}} \right\}^2} = 2.58 \ \mu m$$

For

Rex $= 50, C_D = 1.84$

Xc $= 3.65 \ \mu m.$

VII. COLLISION OF SOLID PARTICLES ON THE SOLID SURFACE BY VOLLHEIM

In order to derive a law of the collision for one solid particle, assuming that one solid particle collides on the solid plane surface, the particle rebounds and rises upward with decreased velocity. The kinetic energy was dissipated by the collision exchanges to heat and to crush energy.[17,18] The process of the energy exchange on the collision may take place by the two independent assumptions:

1. Coefficient K_T of the shock loss by the collision which is related to the kinds of the materials
2. Coefficient f_s of the sliding friction on the collision

A coefficient K_T can be determined by means of the free fall of a solid particle on the solid plane surface, as shown in Figure 19. Then $K_T = 0$ means that material has perfect plasticity and $K_T = 1$ means that material has perfect elasticity. The coefficient of sliding friction can be determined by the force components Fx and Fy as $f_s = Fx/Fy$. The numerical examples of material pairs are shown in Table 1. Those values K_T and f_s depend on the collision velocity, the roughness of the solid surface, and the characters of the materials.

Here, with adapting the relationship of the impulse-momentum equations, the mechanics of the collision mechanism can be described as follows. As shown in Figure 20, before colliding against the solid surface, one solid particle has the velocity component Ux1, Uy1, and Ω1 (angular velocity). Then after colliding against the solid surface, this particle has Ux2, Uy2 and Ω2. Therefore the differences of *U* (velocity) and Ω (angular velocity) can be written

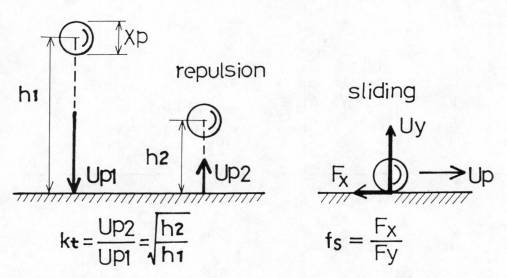

FIGURE 19. Illustration of the determination of coefficients of shock loss and of sliding friction.

Table 1
NUMERICAL EXAMPLES OF
MATERIAL PAIRS

Material pairs	K_T	f_s
Wood and wood	0.50	—
Cork and cork	0.55	—
Lead and lead	0.20	—
Cast iron and cast iron	0.65	—
Wheat and steel	0.50	0.40
Glass and steel	0.94	0.55
Steel and steel	0.55	0.15
Glass and glass	0.95	—

$$\Delta Ux = Ux2 - Ux1 \tag{39}$$

$$\Delta Uy = Uy2 - Uy1 \tag{40}$$

$$\Delta\Omega = \Omega2 - \Omega1 \tag{41}$$

In the case of the sliding collision, we can obtain a relationship between ΔUx and ΔUy from the momentum equation

$$\frac{Fx}{Fy} = \frac{mp \cdot \dfrac{\Delta Ux}{\Delta t}}{mp \cdot \dfrac{\Delta Uy}{\Delta t}} = \frac{\Delta Ux}{\Delta Uy} = f_s \tag{42}$$

According to the condition of the adhesion to the solid surface as

$$\frac{Xp \cdot \Omega2}{2} + Ux2 = 0 \tag{43}$$

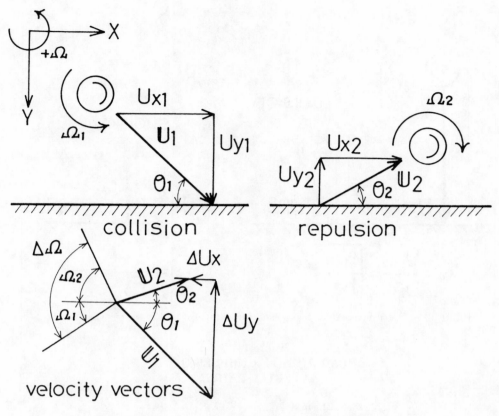

FIGURE 20.   Illustration of the collision mechanism.

we can obtain an equation of $\Omega 2$

$$\Omega 2 = -\frac{2 \cdot Ux2}{Xp} \tag{44}$$

Now from a relationship between the movement of inertia $(mp \cdot Xp^2/10)$ of a solid particle and the angular momentum

$$\left(\frac{1}{10}\, mp \cdot Xp^2\right) \cdot \frac{\Delta\Omega}{\Delta t} = \frac{Xp}{2} \cdot mp \cdot \frac{\Delta Ux}{\Delta t} \tag{45}$$

we can obtain an equation of $\Delta\Omega$

$$\Delta\Omega = \frac{5}{Xp} \cdot \Delta Ux \tag{46}$$

Considering the direction of the motion of the solid particle and adapting a relationship between the collision and the repulsion on the solid surface, we can obtain the following equation

$$k_T = -\frac{Uy2}{Uy1} \tag{47}$$

Substituting Equation 47 into Equation 40, we can obtain an equation of ΔUy (sliding collision)

$$\Delta Uy = -(1 + k_T) \cdot Uy1 \tag{48}$$

Then substituting Equation 48 into Equation 42, we can obtain an equation of ΔUx (sliding collision)

$$\Delta Ux = f_s \cdot \Delta Uy = -(1 + k_T) \cdot f_s \cdot Uy1 \tag{49}$$

Substituting Equation 49 into Equation 46, we can obtain an equation of $\Delta\Omega$ (sliding collision)

$$\Delta\Omega = \frac{5}{Xp} \cdot (-1)(1 + k_T) \cdot f_s \cdot Uy1 = -5(1 + k_T) \cdot f_s \cdot \frac{Uy1}{Xp} \tag{50}$$

On the other hand, substituting Equations 44 and 45 into Equation 41

$$\frac{5}{Xp} \cdot \Delta Ux = -\frac{2\,Ux2}{Xp} - \Omega 1 \tag{51}$$

so we can obtain an equation of Ux2

$$Ux2 = -\frac{5}{2} \cdot \Delta Ux - \frac{Xp \cdot \Omega 1}{2} \tag{52}$$

Here substituting Equation 52 into Equation 39, we can obtain an equation of ΔUx (adhesive collision)

$$\Delta Ux = -\frac{2}{7} \cdot \left(\frac{Xp \cdot \Omega 1}{2} + Ux1 \right) \tag{53}$$

Substituting Equation 53 into Equation 46, we can obtain an equation of $\Delta\Omega$ (adhesive collision)

$$\Delta\Omega = -\frac{5}{7} \cdot \left(\frac{2 \cdot Ux1}{Xp} + \Omega 1 \right) \tag{54}$$

Then transforming Equations 53 and 48, we can obtain an equation of the critical angle $\theta 1K$ between adhesion and sliding on the solid surface

$$\tan \Theta 1k = \frac{Uy1}{Ux1} = \frac{2}{7} \cdot \frac{1 + \dfrac{Xp \cdot \Omega 1}{2 \cdot Ux1}}{1 + k_T} \cdot \frac{1}{f_s} \tag{55}$$

Therefore, a condition of sliding on the solid surface must satisfy the following relation

$$\tan\Theta 1 < \tan\Theta 1k \tag{56}$$

Here, we try to calculate the critical angle $\theta 1K$ under the condition of glass and steel. Therefore the values of K_T and f_s are 0.94 and 0.55. Assuming that a particle diameter Xp is 2 mm, Ux1 is 10 m/s, and angular velocity $\Omega 1$ of a particle is varied, Equation 55 becomes

$$\tan \Theta 1k = \frac{2}{7} \cdot \frac{1 + \dfrac{2 \times \Omega 1}{2000}}{1 + 0.94} \cdot \frac{1}{0.55} = 0.268 \left(1 + \frac{\Omega 1}{10,000} \right)$$

Table 2
NUMERICAL EXAMPLES OF THE CRITICAL
ANGLE θ1K

$\Omega 1$ (rad/s)	0	2π	20π	200π	2000π
tanθ1K	0.268	0.268	0.269	0.285	0.436
θ1K (rad)	0.262	0.262	0.263	0.278	0.411

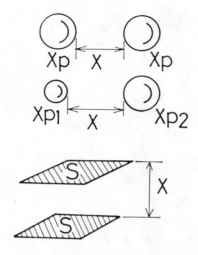

FIGURE 21.   Hamaker model of adhesive force.

The critical angle θ1K is shown in Table 2.

To briefly describe the van der Waals force, at a contact point of the solid particles, an attractive force between molecules in the solid particles acts on each other. Consequently, a combine force is brought about. Therefore, when the solid particles sufficiently approach each other, the solid particles adhere to each other.[19] This combine force F can be estimated by Hamaker, as shown in Figure 21

$$\text{for } X < 1000 \text{ Å}$$

$$\text{spheres } F = \frac{A \cdot Xp}{24 \cdot X^2}$$

$$\text{parallel plates } F = \frac{A \cdot S}{6 \cdot \pi \cdot X^3}$$

and in case of the diameter Xp1 and Xp2 of the spherical particles

$$F = \frac{A}{12 \cdot X^2} \cdot \frac{Xp1 \, Xp2}{Xp1 + Xp2}$$

From a theoretical point of view, the value of A for vacuum is 10^{-19} J, but from the experimental results the values of A are between 10^{-18} to 10^{-21} J. For example, if the diameter of the solid particles is 15 μm, the attractive force F for X = 10 Å becomes

$$F = \frac{10^{-12} \times 15 \times 10^{-4}}{24 \times (10^{-7})^2} = 6.25 \times 10^{-3} \text{ dyn.}$$

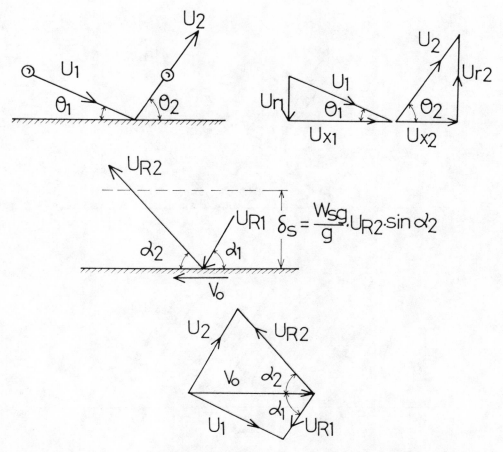

FIGURE 22. Velocity vectors in the case of the collision of solid particles on a plane surface.

VIII. MOTION OF FINE SOLID PARTICLES IN TURBULENT AIR FLOW NEAR A SOLID PLANE SURFACE (BOHNET THEORY)

In order to make clear the motion of the fine solid particles in turbulent air flow near the solid plane surface, Bohnet (1963) published the following idea.[20,21] As shown in Figure 22, a solid particle having an absolute velocity U1 collides on the plane surface with an angle θ1 for a horizontal plane, and then after repulsion a solid particle has an absolute velocity U2 with an angle θ2 for a horizontal plane.[22]

Considering this physical idea more clearly, a particle motion relative air flow of velocity Vo is taken into account. So a particle with a relative velocity UR1 collides on surface with an angle α1 for a horizontal plane and then after repulsion a solid particle has an relative velocity UR2 with an angle α2 for a horizontal plane. Then after a repulsion, a solid particle reaches the vertical distance δs which is independent on a boundary layer of flow.

As shown in Figure 23, a solid particle with mass Ms during the deceleration in the distance δs furnishes its energy to the particle-laden gas. Then a volume $\bar{V}$m of particle-laden gas is accelerated by the momentum exchange and this particle-laden gas escapes with velocity $\bar{U}$R2 from this layer δs. Denoting that the density of gas is ρ and the mixture ratio is μ_s, so an equation of the momentum exchange can be obtained

$$Ms \cdot UR2 = \bar{V}m \cdot \rho \cdot (1 + \mu_s) \cdot \bar{U}R2 \qquad (57)$$

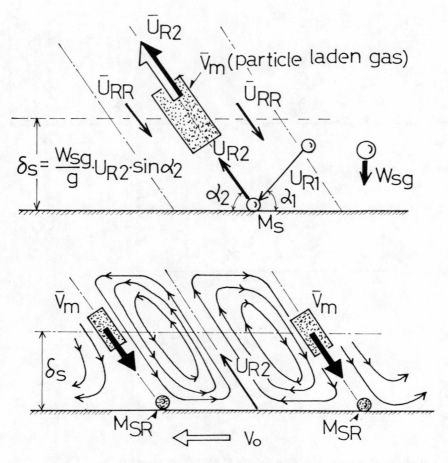

FIGURE 23.   Illustration of the momentum exchange between solid particles and the air flow.

Here, assuming that the motion of the accelerating particle-laden gas is proportional to the particle mass in the volume $\bar{V}m$, a relationship between $UR2$ and $\bar{U}R2$ can be obtained as a proportional factor k. From a condition of the continuity, a volume $\bar{V}$ at the edge of layer δs must return again and, at the same time, particle mass $\bar{V}m{\cdot}\rho{\cdot}\mu_s$ also must return again. The returning velocity at the layer δs denotes $\bar{U}RR$. As a result of the inertial force on the particle-laden gas, a portion of the particles collide on a plane surface as

$$Ms R = \bar{V}m{\cdot}\rho{\cdot}\mu_s{\cdot}\epsilon \qquad (58)$$

Therefore, the secondary flow patterns near the plane surface occurs by the momentum exchange of the solid-particle laden gas flow, as shown in Figure 23. Then in combination with Equation 57 and Equation 58, a following equation can be obtained

$$\frac{Ms - Ms R}{Ms} = \frac{\Delta Ms}{Ms} = 1 - \frac{\bar{V}m{\cdot}\rho{\cdot}\mu_s{\cdot}\epsilon}{\dfrac{\bar{V}m{\cdot}\rho{\cdot}(1 + \mu_s){\cdot}\bar{U}R2}{UR2}} =$$

$$= 1 - \frac{\mu_s}{1 + \mu_s}{\cdot}\frac{UR2}{\bar{U}R2}{\cdot}\epsilon = 1 - \frac{\mu_s}{1 + \mu_s}{\cdot}k{\cdot}\epsilon \qquad (59)$$

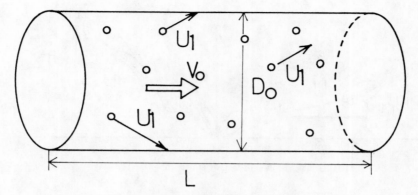

FIGURE 24.   Solid particle laden gas flow in pipe.

If there is an equilibrium state between the collision particle mass on the surface and the repulsion particle mass on the surface, Equation 59 becomes $\Delta Ms = 0$ and an equation for the maximum possible particle laden state can be obtained as

$$\mu_{s \cdot max} = \frac{1}{\epsilon \cdot k - 1} \tag{60}$$

Now a relationship between Vo and $\overline{U}RR$ can be obtained from the momentum-loss of the particles by the repulsion on the surface. As shown in Figure 24, the collision particle mass per unit time is

$$Mss = \pi \cdot Do \cdot L \cdot \rho \cdot \mu_s \cdot UR1 \tag{61}$$

The particle mass Mss accepts the momentum loss by the friction of the particles on the surface as

$$\Delta I = Mss \cdot (Ux1 - Ux2) \approx Mss \cdot \Delta Ux \tag{62}$$

If the friction stress between the solid particle and the surface is proportional to the inertial force of the solid particles, so the frictional force per distance L can be written

$$\Delta Ff = \frac{\mu_s \cdot Vo \cdot \rho \cdot (\pi Do^2/4) \cdot L}{U1} \cdot \frac{U1^2}{2} \cdot \frac{\lambda_z^*}{Do} \tag{63}$$

Then a relationship between $\overline{U}R1$ and Vo can be obtained from a condition of $\Delta I = \Delta Ff$ as

$$\frac{UR1}{Vo} = \frac{\lambda_z^*}{8} \cdot \frac{U1}{\Delta Ux} \tag{64}$$

Now assuming that U1 is nearly equal to Vo, then Equation 64 becomes

$$\frac{UR1}{Vo} = \frac{\lambda_z^*}{8} \cdot \frac{Vo}{\Delta Ux} \tag{65}$$

Therefore, URR/Vo is a function of the friction factor λ_z^* and of $Vo/\Delta Ux$.

In order to estimate the value of $UR1/Vo$ for wheat, the experimental results of Mus-

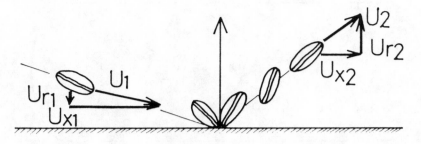

FIGURE 25.   Motion of the wheat near the plane surface.

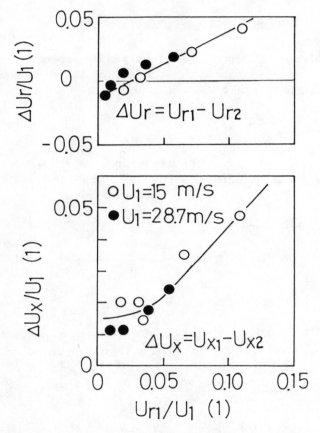

FIGURE 26.   Measured particle (wheat) velocity of diameter X_p = 3 mm for repulsion on an aluminum plate.

chelknautz (1959)[22] can be applied. As shown in Figures 25 and 26 for wheat particles of diameter $X_p \fallingdotseq$ 3 mm on the aluminum plane, the values of $\Delta U_r/U_1$ and $\Delta U_x/U_1$ are nearly 0.025 and 0.033 for U_{r1}/U_1 = 0.075 and U_1 = 15 to 28.7 m/s. Therefore the value of UR_1/V_o for λ_z^* = 0.0032 can be calculated as

$$\frac{UR_1}{V_o} = \frac{0.0032}{8} \times \frac{1}{0.033} \fallingdotseq 0.0121$$

So the relative velocity UR_1 is equal to nearly 1.2% of V_o.

REFERENCES

1. **Hackeschmidt, M.,** *Grundlagen der Strömungstechnik,* Vol. 1, VEB Deutscher Verlag, Leipzig, 1969.
2. **Soo, S. L., Ihrig, H. K., and Kouh, A. F.,** Experimental determination of statistical properties of two-phase turbulent motion, *J. Basic Eng.,* 609, 1960.
3. **Hinze, J. O.,** *Turbulence,* McGraw-Hill, New York, 1959.
4. **Hinze, J. O.,** Turbulent Fluid and Particle Interaction, *Progress in Heat and Mass Transfer,* Vol. 6, 433, 1972.
5. **Lee, N. and Dukler, A. E.,** A stochastic model for turbulent diffusion of particles or droplets, *AICh E. J.,* 27(4), 552, 1981.
6. **Batchelor, G. K.,** Kolmogoroff's theory of locally isotropic turbulence, *Proc. Cambridge Philos. Soc.,* 43, 533, 1947.
7. **Corrsin, S.,** Simple theory of an idealized turbulent mixer, *AIChE. J.,* 3(3), 329, 1957.
8. **Watzel, G.,** Untersuchung von Tropfenbahnen in umgelenkten Strömungen and ihre Anwendung auf die Tropfenabscheidung in Trocknern, *VDI-Forschung.,* 1970.
9. **Hinze, J. O.,** Experimental investigation on secondary currents in the turbulent flow through a straight conduit, *Appl. Sci. Res.,* 28, 453, 1973.
10. **Diukelacker, M., Hessel, M., Meier, G. E. A., and Schewe, G.,** Further results on wall pressure fluctuations in turbulent flow, Proc. 3rd U.S. Fed. Rep. Germany, Munich, 1957.
11. **Rotta, J. C.,** Eine theoritische Untersuchung über den Einfluβ der Druckscherkorrelationen auf die Entwicklung dreidimensionaler turbulente Grenzschichten, *Dtsch. Forschungs. Versuchsanstalt Luft Raumfaurt FB,* 75-05, 1979.
12. **Rumpf, H.,** Versuche zur Bestimmung der Teilchenbewegung in Gasstrahlen und des Beauspruchungsmechanismus in Strahlmühlen, *Chem. Ing. Tech.,* 32(5), 335, 1960.
13. **Laufer, J.,** *NACA-Tech. Note.,* 2954, 1953.
14. **Townsend, A. A.,** *The Structure of Turbulent Shear Stress,* 2nd ed. Cambridge University Press, Cambridge, 1976.
15. **Gluth, M.,** Untersuchungen zur Wurfgebläseförderung, *VDI-Forschung.,* 544, 1971.
16. **Bauman, V. A., Klushantsev, B. V., and Martynov, V. D.,** *Mehanitcheskoe Oborudovanie,* Moskva, 1981, 167.
17. **Vollheim, R.,** *Pneumatischer Transport,* VEB-Deutscher Verlag, Leipzig, 1971.
18. **Housner, G. W. and Hudson, D. E.,** *Applied Mechanics, Dynamics,* D. Van Nostrand, New York, 1950.
19. **Zimon, A. D.,** *Adhesion of Dust and Powder,* (Translation ed. by M. Corn) Plenum Press, New York, 1969.
20. **Bohnet, M.,** Das Verhalten feiner Staubteilchen in strömenden Gasen in Wandnähe, *Z. Angew. Math. Mech.,* 43(1/2), T116, 1963.
21. **Bohnet, M.,** Experimentelle und theoretische Untersuchungen über das Absetzen, das Aufwirbeln und den Transport feiner Staubteilchen in pneumatischen Förderleitungen, *VDI-Forschung.,* 507, 1965.
22. **Muschelknautz, E.,** Theoretische und experimentelle Untersuchungen über die Druckverluste pneumatischer Förderleitungen unte unter besonderer Berücksichtigung des Einflusses von Gutreichung und Gutgewicht, *VDI-Forschung.,* 476, 1959.

Chapter 5

DYNAMICAL SIMILARITY OF THE SOLID PARTICLE

I. INTRODUCTION

When we predict the behavior of the solid particles in a turbulent gas flow, or when we design dust collectors for separation of solid particles from a gas flow, it is valuable to consider the applications of the theory of the mechanical similarity for a solid particle and fluid flow. Therefore, we will consider the derivative process of the mechanical similarity.[1,2]

The necessary conditions for the similarity of the motion of two kinds of dust-particle systems in a fluid are contributed mainly by three terms:

1. Geometrical similarity of the boundary of a fluid flow, for example, construction of the cyclones
2. Similarity of fluid flow in each system
3. Similarity of the loci of dust particles

II. EQUATION OF MOTION AND SIMILARITY FOR THE MOTION OF A SOLID PARTICLE

A. Geometrical Similarity of the Boundary

Figure 1 is an example of cyclone dust collectors, the representative lengths for system I and system II are denoted by $D\acute{o}$ (diameter of inlet pipe), D1 (diameter of cyclone), D2 (diameter of outlet-pipe), and $D\acute{o}$, $D\acute{1}$, $D\acute{2}$, respectively.

Then the condition of the geometrical similarity for the both cases becomes

$$Sl = \frac{Do}{D\acute{o}} = \frac{D1}{D\acute{1}} = \frac{D2}{D\acute{2}} \tag{1}$$

where Sl means a dimensionless number of the geometrical similarity.

B. Similarity of the Fluid Flow

As shown in Figure 1, where Vo is a mean inlet velocity of gas in the inlet pipe, $V\theta$ is a tangential velocity of gas flow, and Vr is a radial velocity of gas flow, then the condition of the similarity of fluid flow must be satisfied for systems I and II as follows

$$Sv = \frac{Vo}{V\acute{o}} = \frac{V\theta}{V\acute{\theta}} = \frac{Vr}{V\acute{r}} \tag{2}$$

where Sv is a dimensionless number of similarity of the fluid flow.

C. Similarity of the Loci of Dust Particles

The tangential and radial velocities of dust particles for systems I and II are $U\theta$, Ur, and $U\theta'$, Ur', respectively. Then in order to satisfy the similarity of the loci of dust particles, a following condition is necessary

$$Su = \frac{Uo}{U\acute{\theta}} = \frac{Ur}{Ur'} \tag{3}$$

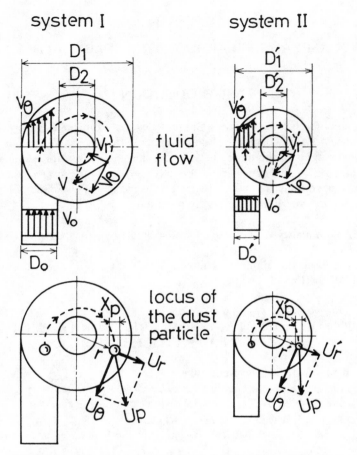

FIGURE 1.   Illustrations of the dynamical similarity of the solid particle in a separation chamber.

where Su is a dimensionless number for the loci of dust particles. In addition to this, a condition of similarity of time lapse t and t′ for the systems I and II must be satisfied

$$St = \frac{t}{t'} \tag{4}$$

where St is a dimensionless number of the lapse of time. Therefore, a dimensionless number Su can be transformed to

$$Su = \frac{Ur}{Ur'} = \frac{(dr/dt)}{(dr'/dt')} = \frac{(dr/dr')}{(dt/dt')} = \frac{Sl}{St} \tag{5}$$

where r and r′ are the radial positions of dust particles in systems I and II, respectively.

Consequently, a relationship of the dimensionless number between Sl, St, and Su must be satisfied from Equation 5 as

$$Sl = St \cdot Su \tag{6}$$

Now introducing a similarity of acceleration of dust particles in dust collectors, the following equation must be satisfied

$$Sa = \frac{\left(\dfrac{dUr}{dt}\right)}{\left(\dfrac{dUr'}{dt'}\right)} = \frac{\left(\dfrac{dUr}{dUr'}\right)}{\left(\dfrac{dt}{dt'}\right)} = \frac{Su}{St} \tag{7}$$

where Sa is a dimensionless number of the similarity of dust particle acceleration. Therefore, a relationship of the dimensionless numbers between Sa, St, and Su becomes

$$Su = Sa \cdot St \tag{8}$$

On the other hand, from the hydrodynamical point of view, one of the most important conditions of dynamical similarities concerning the similarity of fluid flow for both geometrical similarity systems must take the same value of Reynolds number Re, which is a ratio of the inertial force of the fluid particle to the viscous force of the fluid particle for the both systems I and II as

$$Re = \frac{D1 \cdot Vo}{v} = \frac{D'1 \cdot V'o}{v'} = const. \tag{9}$$

where v and v' are the kinematic viscosity of gas for both systems. Then, assuming that the drag force acting on a solid particle of diameter Xp obeys the Stokes resistance force, which means that the drag force is proportional to the relative velocity between the fluid velocity V and particle velocity Up, the equation of motion of a solid particle for the radial direction can be written

$$\rho_p \cdot \frac{\pi \cdot X_p^3}{6} \cdot \frac{dUr}{dt} = \rho_p \cdot \frac{\pi \cdot X_p^3}{6} \cdot \frac{U_\theta^2}{r} - 3 \cdot \pi \cdot \eta \cdot Xp \cdot (Ur - Vr), \tag{10}$$

Transforming Equation 10, the following equation can be obtained

$$\frac{dUr}{dt} = \frac{U_\theta^2}{r} - \frac{18 \cdot \eta}{\rho_p \cdot Xp^2} \cdot (Ur - Vr) \tag{11}$$

Here, in order to obtain the law of dynamical similarity for both systems I and II, the following dimensionless quantities and τ must be introduced

$$Ur = Su \cdot U'r, \, U\theta = Su \cdot U'\theta, \, Vr = Sv \cdot V'r, \, t = St \cdot t',$$

$$\tau = \frac{\rho_p \cdot X_p^2}{18 \cdot \eta} \div \frac{Wsg}{g} \tag{12}$$

where Wsg is the terminal velocity of a solid particle in quiet gas. Therefore, the radial acceleration dUr/dt of the solid particle can be transformed to

$$\frac{dUr}{dt} = \frac{d(Su \cdot U'r)}{d(St \cdot t')} = \frac{Su}{St} \cdot \frac{dU'r}{dt'} = Sa \cdot \frac{dU'r}{dt'} \tag{13}$$

Also other quantities of Equation 11 can be transformed as follows

$$\frac{U\theta^2}{r} = Sa \cdot \frac{U'^2_\theta}{r'} \tag{14}$$

$$\frac{18 \cdot \eta}{\rho_p \cdot Xp^2} \cdot (Ur - Vr) = \frac{Sv \cdot V'r}{\tau} \cdot \left(\frac{Su \cdot U'r}{Sv \cdot V'r} - 1 \right) \tag{15}$$

Substituting Equations 13, 14, and 15 into Equation 11, Equation 11 becomes

$$\frac{dU'r}{dt'} = \frac{U'^2_\theta}{r'} - \frac{Sv \cdot V'r}{\tau \cdot Sa} \cdot \left(\frac{Su \cdot U'r}{Sv \cdot V'r} - 1 \right) \tag{16}$$

Then the motion of the equation of a dust particle for the radial direction on second system II can be written

$$\frac{dU_r'}{dt'} = \frac{U_\theta^2}{r'} - \frac{1}{\tau'} \cdot (U_r' - V_r') \tag{17}$$

Now, transforming the second of the right hand terms, Equation 17 becomes

$$\frac{dU_r'}{dt'} = \frac{U_\theta^2}{r'} - \frac{V_r'}{\tau'} \cdot \left(\frac{U_r'}{V_r'} - 1 \right) \tag{18}$$

Consequently, concerning the conditions of similarity of the dust particle motion, we can obtain the following equations from Equations 16 and 18

$$\frac{V_r' \cdot Sv}{\tau \cdot Sa} = \frac{V_r'}{\tau'}, \quad \frac{Su \cdot U_r'}{Sv \cdot V_r'} = \frac{U_r'}{V_r'} \tag{19}$$

Therefore, Equation 19 becomes

$$\tau' = \tau \cdot \frac{Sa}{Sv} = \frac{\tau}{St}, \quad Su = Sv \tag{20}$$

Here, a similarity number St can be transformed as

$$St = \frac{Sl}{Su} = \frac{Sl}{Sv} = \frac{D1}{D1'} \cdot \frac{V_o'}{Vo} \tag{21}$$

so substituting Equation 21 into τ' of Equation 20, we can obtain the following equation

$$\frac{\tau \cdot Vo}{D1} = \frac{\tau' \cdot V_o'}{D1'} = Stk = \text{const.} \tag{22}$$

Finally, transforming τ and τ' into the original equations, the following transformed equation can be obtained:

$$\frac{Vo \cdot Wsg}{D1 \cdot g} = \frac{V_o' \cdot Ws_g'}{D1' \cdot g} = Sst = \text{const.} \tag{23}$$

In general, the dimensionless number Sst is called the Stokes number or inertia parameter. Stokes number is one of the most important dimensionless numbers for investigating the behavior of dust particle motion in a fluid flow. The physical meaning of Stokes number or inertia parameter, is a ratio of inertial force to viscous force for the dust particle, or a ratio of a representative length D1 of the system to the stop-distance of the dust particle ejecting with initial velocity Vo into the quiet gas under the assumption of the Stokes drag force.

On the contrary, as a condition of similarity for considering the influence of the gravity force of the dust particle, the following condition must be satisfied as

$$\frac{U_\theta^2}{r} = \frac{U_\theta^2}{r'} \tag{24}$$

Then, taking the representative lengths D1 and D1' and the representative velocities Vo and Vó instead of $U\theta$ and $U\theta'$, we can obtain the following equation

$$\frac{V_o^2}{D1 \cdot g} = \frac{V_o'^2}{D1' \cdot g} = Fr = \text{const.} \tag{25}$$

This dimensionless number Fr is called a Froude number, which means the ratio of dust particle inertial force to the gravity force of the dust particle.

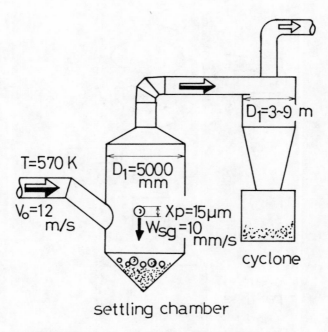

FIGURE 2.   Dust settling chamber for the furnace system.

To sum up the above stated results, the conditions of motion similarity of dust particles for both systems I and II should be considered on the basis of the equivalences of the Reynolds number and Stokes (inertial) number. In addition to this, in considering the influence of the gravity force, the equivalence of the Froude number must be considered. Further, in a special case, taking in account gravity force, we must consider a new dimensionless number by the relations of Stokes number and Froude number as

$$\frac{Stk}{Fr} = \frac{D1 \cdot g}{Vo^2} \cdot \frac{Vo \cdot Wsg}{D1 \cdot g} = \frac{Wsg}{Vo} = \text{const.} \tag{26}$$

III. EXPERIMENTS OF THE SIMILARITY BY THE MODELS OF SETTLING CHAMBERS (BARTH EXPERIMENTS)

In model experiments, to grasp the separation process of dust particles based upon the mechanical principle, it is necessary to consider the equivalence of the similarities, namely, not only the fluid flow in the dust collector, but also the motion of the dust particle.[3] This principle can be applied to the problem of the motion of river sand.[4]

Here, the dust-laden gas is transported to the cyclone dust collector of diameter D1 = 3 to 9 m through the dust settling chamber in which the coarse dust is sedimented to the chamber on the furnace system as shown schematically in Figure 2. In this system, in order to estimate the dust collection in the settling chamber, Barth did experiments using three kinds of settling chamber models, as shown in Figure 3. The numerical values are shown in Table 1.

A. Model I

Model I (D1 = 250 mm) uses water as the fluid. Its size is 1/20 of the original settling chamber. The similarity conditions of the flow Reynolds number and the flow Froude number are equivalent between the original and model I. In model I, even if the size of the chamber is 1/20 of the original, kinematic viscosity υ (m²/s) of water is lower than that of air. When two kinds of similarity conditions, D1/Xp and ρ_p/ρ, must be satisfied for the original and

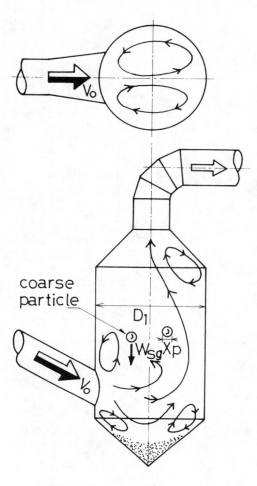

FIGURE 3.   Settling chamber model.

model I, the density of dust ρ_p and the particle size Xp for the model I must be 4×10^6 kg/m^3 and 0.75 µm, respectively.

However, it is very difficult to consider such a heavy, fine dust particle, therefore from the practical point of view, one will recognize that the satisfaction of all of the similarity conditions for the model is tremendously difficult. On the other hand, from the equation of the motion for a dust particle, instead of the very heavy, fine dust particle, using the low density and fine dust particle, if the additive similarity condition Wsg/Vo = const for the original and the model I is satisfied, then the similarity loci of dust particles in both chambers will be expected. Here, Vo is a representative velocity of fluid in this chamber. This fact is connected with the following additive similarity conditions by the two assumptions:

1. The drag force acting on the dust particle is obeyed by Stokes drag force
2. The value of $(\rho_p - \rho)/\rho_p$ is nearly equivalent to 1

From those two assumptions, the terminal velocity Wsg of the dust particle by the gravitational force in the quiet gas can be written

$$\text{Wsg} = \frac{\rho_p \cdot g \cdot X_p^2}{18 \cdot \eta} \qquad (27)$$

Table 1
MODEL EXPERIMENTS OF THE SETTLING CHAMBER FOR THE
COARSE DUST PARTICLES BY BARTH

			Model		
		Original	**I**	**II**	**III**
Diameter D_1(mm)		5000	250	500	500
Fluid		Furnace gas	Water	Air	Air
Temperature of gas T(K)		573	324	293	293
Density of fluid ρ (kg/m³)		0.614	987	1.20	1.20
Kinematic viscosity υ (m²/s)		49×10^{-6}	0.547×10^{-6}	15.1×10^{-6}	15.1×10^{-6}
Inlet velocity Vo (m/s)		12.0	2.68	3.80	37.0
Diameter of dust particle Xp (μm)		15.0	38.7	4.70	1.50
Density of dust particle ρ_p (kg/m³)		2500	2500	4870	4870
Terminal velocity Wsg (mm/s)		10.0	2.24	3.17	0.324
Condition of similarity	D_1/Xp	3.33×10^5	6.45×10^3	1.06×10^5	3.33×10^5
	Re(VoD$_1$/υ)	1.22×10^6	1.22×10^6	0.126×10^6	1.22×10^6
	Fr(Vo/$\sqrt{gD_1}$)	1.72	1.72	1.72	16.8
		4070	2.54	4070	4070
Additive condition	Wsg/Vo	8.35×10^{-4}	8.35×10^{-4}	8.35×10^{-4}	0.87×10^{-4}
	WsgVo/gD$_1$	2.44×10^{-3}	2.44×10^{-3}	2.44×10^{-3}	2.44×10^3

Consequently, concerning the dust particle as an additive similarity condition for using the model, when a condition of Wsg/Vo = const is sufficiently satisfied, all of the common similarity conditions 1 through 4, as described in Table 1 will not always be necessary to satisfy; particularly in the case of model I (water), an assumption of $(\rho_p - \rho)/\rho_p$ is not satisfied.

Thus, it is necessary to discuss the possibility of the shift — how the size can be enlarged from the experimental results by the model under the dissatisfaction of an assumption (2) of $(\rho_p - \rho)/\rho_p$. And, in addition to this, when the influence of gravity force for the motion of the dust particle cannot be neglected, the equivalence of the Stokes number Vo·Wsg/D1·g = const as a similarity condition must be satisfied from the equation of motion of the dust particle.

B. Model II

In the case of model II, the size of model is 1/10 of the original. The conditions of the Froude number of flow do satisfy, but those of the Reynolds number do not. In this case, the additive similarity conditions must be introduced under the dissatisfactions of the common similarity conditions 1 and 2 and also under the satisfaction of both assumptions 1 and 2.

C. Model III

The size of model III is 1/10 of the original. The common similarity conditions 1, 2, and 4 without 3 are satisfied, but an additive condition Wsg/Vo is not satisfied in comparison with the original. Then, when the gravity force of the dust particle does not influence the motion of the dust particle, and also when the terminal velocity of this dust particle can be neglected in comparison with the representative velocity of gas in the chamber, the experimental results of this model can be applied to the large scale of the separation chamber.

As is known from these examples, it is very difficult to satisfy all of the similarity conditions as stated above. But from the practical point of view, we can estimate the behavior of dust particles under the recognition of the relative error based upon the experimental results which satisfy only one important similarity condition.

IV. MODEL EXPERIMENT OF THE HYDRAULIC CYCLONE FOR COLLECTION EFFICIENCY BY BARTH AND TRUNZ

When the Reynolds number, the Froude number, and the value of ρ_p/ρ are equal for both systems, namely the original and the model, the separation processes become similar.[5] Instead of the above stated dimensionless numbers, we can apply the following additive similarity conditions which are derived from the equation of motion for the dust particle

$$\frac{Ur}{V} = \text{const.} \tag{28}$$

$$\frac{Ws_g^{2-K} \cdot V^K}{D1 \cdot g} = \text{const.} \tag{29}$$

where the value of the exponent k is equal to 1 for the Stokes flow and equal to 0 for the Newton flow and also Wr is the relative velocity between the dust particle and the fluid flow. Those two additive conditions are satisfied under the following assumptions:

1.   The drag force D on the dust particle can be written as

$$D = \frac{c_D}{Rep} \cdot \frac{\pi \cdot X_p^2}{4} \cdot \frac{\rho \cdot W_r^2}{2} \tag{30}$$

$$Rep = \frac{Xp \cdot Ur}{v} \tag{31}$$

2.   The value of $(\rho_p - \rho)/\rho_p$ is nearly equal to 1.

Then the hydraulic cyclone as shown in Figure 4 was applied for the separation of the various sizes of carborundom (SiC) and with various water quantities. Figure 5 shows the collection efficiency for the water flow rate Qo. Here denoting that Wr2 is a relative fluid velocity for a dust particle rotating with the tangential velocity Uθ2 and with terminal velocity Wsg around an imaginary cylinder A2 of diameter D2, a dynamical relationship between Wr2, Vθ2, Wsg, and D2 can be written

$$Wr2 = Wsg \cdot \left(\frac{2 \cdot V\theta 2^2}{D2 \cdot g}\right)^{\frac{1}{2-K}} \tag{32}$$

Then assuming that the inward radial velocity Vr2 of fluid on the surface of an imaginary cylinder A2 is a uniform distribution, Barth and Trunz considered that the dust particles with a condition Wr2/Vr2 > 1 must be all separated, but those with a condition Wr2/Vr2 < 1 are not separated. Therefore Wr2/Vr2 can be represented with Equation 29 as

$$\left(\frac{Wr2}{Vr2}\right)^{2-K} = 2 \cdot \frac{Wsg^{2-K} \cdot Vr2^K}{D2 \cdot g} \cdot \left(\frac{V\theta 2}{Vo} \cdot \frac{A2}{Ao}\right)^2 \tag{33}$$

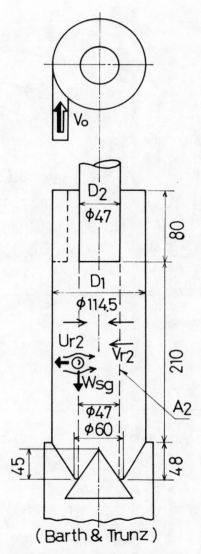

FIGURE 4.   Model of the hydraulic
cyclone.

Consequently, in order to calculate the values of Wr2/Vr2, they assumed $V\theta2/Vo = 1.96$
and $k = 0.8$. The calculated results are summarized in Figure 6. From this figure, you will
find that all of the experimental values fall on a curve of $Wsg/Vr2 \leqq 0.2$.

By the application of this technical method of the similarity, we can estimate the collection
efficiency for the large type of the dust collector, where for the gas collector the value of
exponent k may be used $k \fallingdotseq 1$.

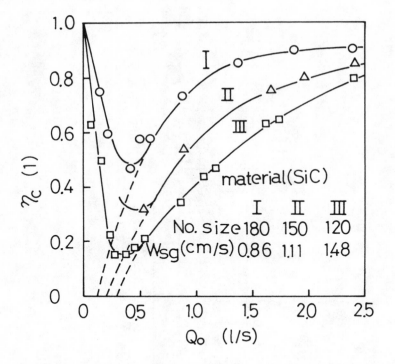

FIGURE 5.   Collection efficiency of the hydraulic cyclone for the water flow rate.

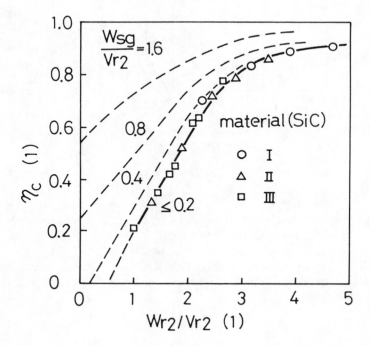

FIGURE 6.   Relationship between the collection efficiency and the flow similarity.

REFERENCES

1. **Fuchs, N. A.,** *The Mechanics of Aerosols,* Pergamon Press, New York, 1964.
2. **Soo, S. L.,** Fluid Dynamics of Multi-Phase Systems, Blaisdell Publishing, 1963.
3. **Barth, W.,** Die Anwendung des Modellversucher zur Lösung strömungstechnischer Aufgaben, *Z. Ver. Dtsch. Ing.,* 92(5), 105, 1950.
4. **Graf, W. H.,** *Hydraulics of Sediment Transport,* McGraw-Hill, New York, 1971.
5. **Barth, W. and Trunz, K.,** Modellversuche mit wasserdurchströmten Zyklonabscheidern zur Vorausbestimmung der Abscheideleistung, *Z. Angew. Math. Mech.,* 3(8/9), 255, 1950.

Chapter 6

FRACTIONAL COLLECTION EFFICIENCY AND TOTAL COLLECTION EFFICIENCY

I. INTRODUCTION

There are many papers concerning the characteristics of total collection efficiency η_c, but rather fractional collection efficiency $\eta_x(Xp)$ for a given dust in relation to the protection against air pollution.

For example, Wicke and Krebs (1971) investigated the characteristics of fractional collection efficiency $\eta_x(Xp)$ of various kinds of scrubbers, i.e., cyclonic deduster, packed-bed spray deduster, plate scrubber, self-induced spray deduster, and Venturi scrubber. Also Lapple and Kamack (1956) investigated the characteristics of fractional collection efficiency of wet-dust scrubbers.

Figure 1 shows the curves of fractional collection efficiency $\eta_x(Xp)$ of cyclone dust collectors. From those curves of $\eta_x(Xp)$, the equation of $\eta_x(Xp)$ can be represented by the following equation

$$\eta_x(Xp) = 1 - \exp(-\alpha \cdot X_p^m) \tag{1}$$

Defining the cut-size Xc which corresponds to the size of $\eta_x(Xp) = 0.5$ (50%), the value of α can be obtained as

$$\alpha = \frac{\ln 2}{Xc^m} = \frac{0.693}{Xc^m} \tag{2}$$

Substituting Equation 2 into Equation 1, we can obtain the general equation of the fractional collection efficiency $\eta_x(Xp)$ as

$$\eta_x(Xp) = 1 - \exp\left\{-\ln 2 \left(\frac{Xp}{Xc}\right)^m\right\} \tag{3}$$

The separation indexes m of Equation 3 for the general types of cyclones become m = 0.8 to 1.5, for the rotary flow dust collectors (Drehströungsentstauber, D.S.E.) m = 1.5 to 2.5, for the spray tower m ≒ 1.5, and the Venturi scrubber[1] m ≒ 2.0.

II. RELATIONSHIP BETWEEN TOTAL COLLECTION EFFICIENCY AND FRACTIONAL COLLECTION EFFICIENCY

Here we consider the relationship between total collection efficiency η_c and fractional collection efficiency $\eta_x(Xp)$, when the size distribution of solid particles can be represented by the cumulative distribution of the Rosin-Rammler-Intelmann equation as follows

$$R(Xp) = \exp(-\beta \cdot Xp^n) \tag{4}$$

where the values of β and n represent the characteristics of the size distribution of the dust particles. When the value of β becomes too large, the size distribution of the dust particles is composed of fine particles.

When the value of n becomes too large, the size distribution of dust particles is composed of nearly the same size distributions. There are many books and many papers concerning the dust size distributions.[2-6]

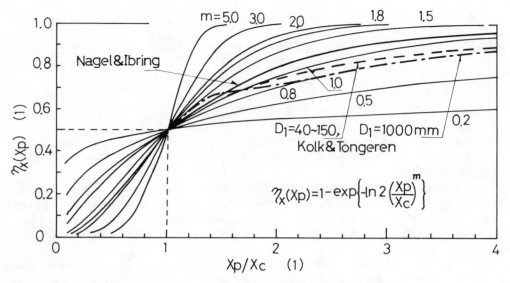

FIGURE 1.   Curves of fractional collection efficiencies.

Generally speaking, the values of β and n are not always constant values within the dust sizes, but change from n_1 and n_2 and also from β_1 to β_2 with an increase in the size of the dust particles. In spite of this phenomenon, from the practical point of view, we represent the size distribution by one approximate Equation 4.

Defining the particle size X_{R50} which corresponds to $R(Xp) = 0.5$ (50%) of Equation 4, the value of β becomes:

$$\beta = \frac{\ln 2}{X^n_{R50}} = \frac{0.693}{X^n_{R50}} \tag{5}$$

Also, the equation of the frequency distribution represented by Equation 4 can be written

$$f(Xp) = -\frac{dR(Xp)}{dXp} = \beta \cdot n \cdot X_p^{n-1} \cdot \exp(-\beta X_p^n) \tag{6}$$

where the unit of f(Xp) is %/μm.

The physical illustrations of Equations 5 and 6 are shown in Figure 2. Therefore, the total collection efficiency η_c can be represented by

$$\eta_c = \int_0^\infty f(Xp) \cdot \eta_x(Xp) \cdot dXp = \int_0^\infty \beta \cdot n \cdot X_p^{n-1} \cdot \exp(-\beta X_p^n) \cdot$$

$$\left[1 - \exp\left\{ -\ln 2 \cdot \left(\frac{Xp}{Xc}\right)^m \right\} \right] \cdot dXp =$$

$$= \int_0^\infty \frac{(\ln 2)}{X_{R50}} \cdot n \cdot \left(\frac{Xp}{X_{R50}}\right)^{n-1} \cdot \exp\left\{ -\ln 2 \cdot \left(\frac{Xp}{X_{R50}}\right)^n \right\} \cdot$$

$$\left[1 - \exp\left\{ -\ln 2 \cdot \left(\frac{Xp}{Xc}\right)^m \right\} \right] dXp \tag{7}$$

In the special case, if we assume that the value of n is equal to m(= Λ), so the total collection efficiency $\eta_{c\Lambda}$ can be written

$$\eta_{c\Lambda} = \frac{1}{1 + (Xc/X_{R50})^\Lambda} \tag{8}$$

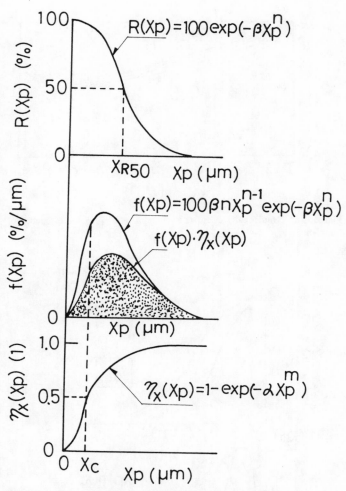

FIGURE 2.   Physical illustrations of the fractional collection efficiency and particle size.

Therefore, for the general types of gas cyclones, the values of m are nearly equal to m $\fallingdotseq$ 1. Assuming that the value of n is equal to 1 (m $\fallingdotseq$ 1), then the total collection efficiency η_c becomes

$$\eta_c = \frac{1}{1 + (X_c/X_{R50})} \tag{9}$$

In the more general case, assuming the values of n = 1 and β = 0.075 (for example, talc powder), the numerically calculated values of η_c of Equation 7 with parameter m = 0.5 to 2.0 are shown in Figure 3.

III. DETERMINATION OF FRACTIONAL COLLECTION EFFICIENCY BY SCHMIDT

As shown in Figure 4, denoting that R_1 (Xp) is the cumulative dust size distribution feeding into the cyclone (dust collector), R_2 (Xp) is the cumulative dust size distribution escaping from the inner pipe of the cyclone (dust collector), and the frequency distribution of R_1 (Xp) and R_2 (Xp) are f_1 (Xp) and f_2 (Xp), respectively, then the total collection

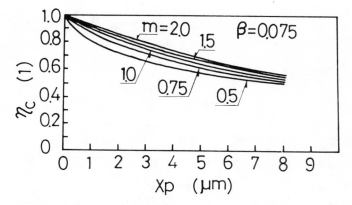

FIGURE 3. Numerically calculated values of total collection efficiency of Equation 7.

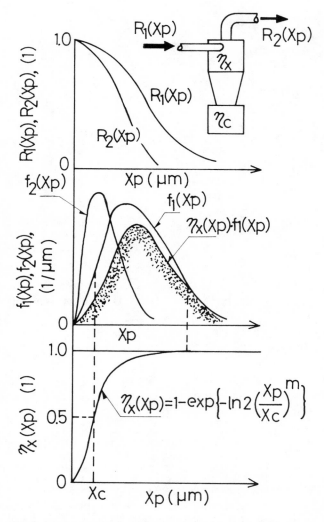

FIGURE 4. Illustration for the determination of the fractional collection efficiency.

efficiency and fractional collection efficiency for the test dust R_1 (Xp) are shown by η_c and η_x (Xp). Now for the fractional collection efficiency f_1 (Xp) of dust size Xp feeding into the cyclone, the dust quantities escaping from the cyclone (dust collector) become

$$f_1(Xp)\cdot(1 - \eta_x(Xp)) \qquad (10)$$

At the same time, for the same dust size Xp, the escaping dust quantities from the cyclone (dust collector) can be represented as

$$f_2(Xp)\cdot(1 - \eta_c) \qquad (11)$$

if we assume that the rate of escaping from the cyclone (dust collector) $(1 - \eta_c)$ is independent of all of the dust sizes Xp escaping from the cyclone (dust collector). Therefore, the value of Equation 10 is equal to the value of Equation 11, and so we can obtain the following equation:

$$f_1(Xp)\cdot(1 - \eta_x(Xp)) = f_2(Xp)\cdot(1 - \eta_c) \qquad (12)$$

Consequently the fractional collection efficiency η_x (Xp) can be represented as

$$\eta_x(Xp) = 1 - \frac{f_2(Xp)}{f_1(Xp)}\cdot(1 - \eta_c) \qquad (13)$$

Still more, using the relations of f_1 (Xp) $= -dR_1$ (Xp)/dXp and f_2 (Xp) $= -dR_2$ (Xp)/dXp, finally Equation 13 becomes

$$\eta_x(Xp) = 1 - \frac{dR_2}{dR_1}\cdot(1 - \eta_c) \qquad (14)$$

This relationship of Equation 14 can be applied to the other types of dust collectors.

From the theoretical point of view, Molerus and Hoffmann (1969)[7,8] derived the equation of the fractional collection efficiency by using the Fokker-Plank equation as follows:

for the Stokes drag,

$$\eta_x(Xp) = \frac{1}{1 + \left(\frac{Xc}{Xp}\right)^2 \cdot \exp\left[T\left\{1 - \left(\frac{Xp}{Xc}\right)^2\right\}\right]} \qquad (15)$$

for the Allen drag,

$$\eta_x(Xp) = \frac{1}{1 + \left(\frac{Xc}{Xp}\right) \cdot \exp\left[T\left\{1 - \left(\frac{Xp}{Xc}\right)\right\}\right]} \qquad (16)$$

where T (T $=$ L Wsc/D) is a dimensionless separation parameter, L is a representative length in the separation chamber, Wsc is the centrifugal sedimentation velocity of dust particles, and D is the coefficient of diffusion. The numerical curves of Equations 15 and 16 are shown in Figures 5 and 6, respectively.

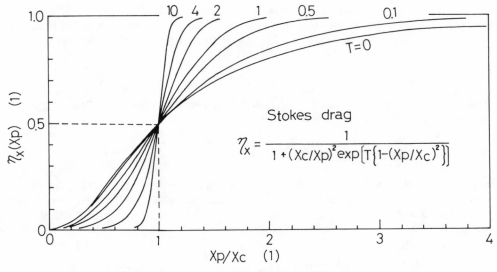

FIGURE 5.   Numerical curves of Equation 15 for Stokes drag.

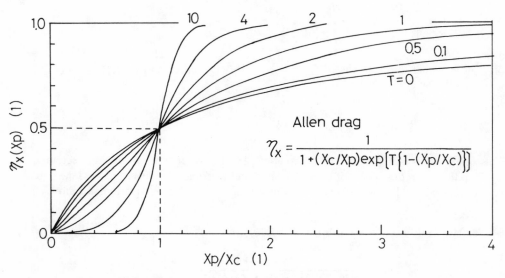

FIGURE 6.   Numerical curves of Equation 16 for Allen drag.

REFERENCES

1. **Hulse, C. and Strauss, W.,** *Choice of Gas Cleaning Plant, Air Pollution,* Thring, M. W., Ed., Butterworths, London, 1957.
2. **Lin, B. Y. H.,** *Fine Particles,* Academic Press, New York, 1976.
3. **Herdan, G. and Smith, M. L.,** *Small Particle Statistics,* Elsevier, Amsterdam, 1953.
4. **Blacktin, S. C.,** *Dust,* Chapman & Hall, London, 1934.
5. **Green, H. L. and Lane, W. R.,** *Particulate Clouds,* E. & F. N. Spon, Ltd., New York, 1964.
6. **Ricci, L.,** *Separation Techniques 2, Gas/Liquid/Solid Systems,* McGraw-Hill, New York, 1980.
7. **Molerus, O. and Hoffmann, H.,** Darstellung von Windsichtertrennkurven durch ein stochastisches Modell, *Chem. Ing. Tech.,* 41(5,6), 340, 1969.
8. **Ogawa, A.,** On the fractional collection efficiencies (in Japanese), *Fluid Eng.,* 12(4), 229, 1976.

Chapter 7

SIMPLE DESCRIPTION OF PNEUMATIC TRANSPORTATION (CHIP TRANSPORTATION)

I. INTRODUCTION

Recently the techniques of pneumatic transportation in many industries, such as chemical engineering, mechanical engineering, food engineering, medicine engineering, and also theories of the pneumatic transportation have been extremely developed. Describing its theoretical developments is beyond the limitations of these volumes. However, it is very important to illustrate the physical mechanism of pneumatic transportation in relation to the separation of the solid particles in a gas stream. Therefore, the author will explain an example of the pneumatic transportation for chips in the wood industry by referring to the book *Vnutrizavodskoj Transport* (1978).[1]

II. CHARACTERISTICS OF TRANSPORTED MATERIALS

Solid particles which are produced by crushing and sawing have the distribution of the particle size Xp. The density of the wood particles, which is a function of humidity and type, are in general $\rho_s = 400$ to 1000 kg/m^3. The relative humidity of wood particles will vary between 8 and 50%. Crushed wood particles have the form of a piece of board. Sawdust has the form of a parallel pipette. The sizes of crushed wood particles do not exceed about Xp = 5.5 mm, but the sizes of the shaved wood vary between Xp = 0.15 mm and Xp = 1.5 mm.

One of the most important factors for the aerodynamics of wood particles is the terminal velocity Wsg which represents their behavior in turbulent air flow. Then, if the terminal velocity Wsg of wood particles in the vertical pipe is equal to the upward velocity of gas, the wood particles gain a state of fluidization in the vertical pipe.

Morikawa[2-4] proposed one classification for the pneumatic transportation of the solid particles, as shown in Table 1. Recently there have been many practical examples and fundamental investigations concerning the low velocity and high concentration pneumatic transportation in Japan and abroad. Table 1 shows a method of judging and considering systematic designs of the pneumatic transportation equipment. Therefore, denoting that the weight of a wood particle is G(N), the mixture of wood particles is in a state of mechanical equilibrium with the pressure force D(N) of the vertical upward air flow. Assuming that the vertical upward air velocity is equal to Wsg, the state of mechanical equilibrium can be written

$$G = D \tag{1}$$

where D can be written

$$D = C_D \cdot \frac{A \rho W_{sg}^2}{2} \tag{2}$$

Here C_D is the drag coefficient of a solid particle which is determined by the flow Reynolds number, the surface roughness, and the shape factor of the solid particle, A is a projected area on a perpendicular plane to the flow direction of the air flow and $\rho W^2sg/2$ is a dynamic pressure of air flow.

Table 1
TYPICAL PNEUMATIC TRANSPORTATION OF SOLID PARTICLES FOR PIPE DIAMETER D = 100 MM BY MORIKAWA

Transportation system		Velocity (m/s)		Mixture ratio M (1)	Pressure drop /100 m pipe length Δp (Pa)
		Air Vo	Particle Up		
A	Rough particle	20—35	(0.5—0.8) Vo	<30	0.1—1 × 10⁵
	Fine particle	15—30	(0.9—1.0) Vo	<10	0.1—1 × 10⁵
B	Fluidlift (Upu fine particle, Upb bottom part)	10—25	Upper part (0.9—1.0)Vo Bottom part (0.1—0.3) Vo	10—30	1.0—3.0 × 10⁵
	Unstable Flow (Upu, Upb)	5—15	Upper part (0.9—1.0) Vo Bottom part ≒ 0.7 Vo	10—30	———
C	For short length of pipe line	3—10	≒Vo	≒150	≒1.0 × 10⁵
	Fluidstat Xp ≤ 2 mm slug-phase transport	3—15	(0.5—0.8) Vo	20—150	0.5—0.6 × 10⁵
D		0.5—8.0	≒Vo	≒150	≥6.0 × 10⁵
	Pulse-phase transport	2—10	≒0.6 Vo	≒150	0.5—6.0 × 10⁵

Now denoting that the symbols l, b, and h are width, side, and thickness of a solid particle, respectively, as shown in Figure 1, Equation 1 can be written

$$C_D \, b \, l \frac{\rho \, W_{sg}^2}{2} = b \cdot h \cdot l \cdot \rho_s \cdot g \tag{3}$$

Therefore, the terminal velocity Wsg can be represented as

$$W_{sg} = 4.43 \sqrt{\frac{h \cdot \rho_s}{C_D \cdot \rho}} \tag{4}$$

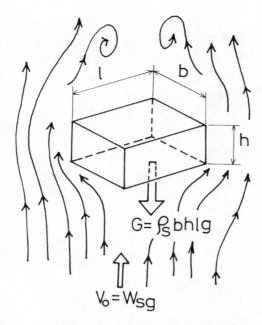

FIGURE 1.   Solid particle and gas flow.

From the experimental results of the drag force D for the wood particles, the drag coefficient C_D could be described as

$$C_D = 0.02\,h + \phi \tag{5}$$

where if a symbol $\underline{h}$ represents by the unit mm and also gas density ρ is 1.2 kg/m³, then Equation 4 may be represented as

$$Wsg = 0.128 \sqrt{\frac{\rho_s}{0.02 + (\phi/h)}} \tag{6}$$

Here, a coefficient ϕ is a form function of solid particles. For example, ϕ is 0.9 for a particle forming a thin plate and $\phi = 1.1$ for particle forms which are nearly equal, b = h, or nearly round. Therefore from Equation 6, we find that the terminal velocity Wsg is influenced by the density ρ_s and the thickness h of the wood particle, as shown in Figure 2.

From the technical point of view, the terminal velocity Wsg is related to the wood particle density and size varies between Wsg = 5 m/s and Wsg = 14 m/s. On the other hand, Table 2 shows spherical and irregular particles which are equivalent to the same terminal velocity of a diameter Xp of a spherical particle corresponding to the particle Reynolds number $10^{-3} \leqq Rex \leqq 10$.

III. EFFECTS OF CONCENTRATION (MIXTURE RATIO)

A mixture ratio M is one of the most important factors for the pneumatic transportation through the pipe line. The mixture ratio M (1) can be represented by the ratios of mass flux

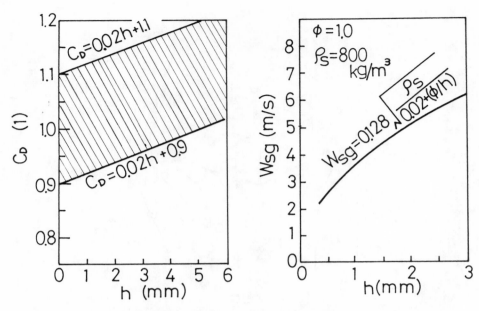

FIGURE 2.   Drag coefficient C_D and the terminal velocity Wsg for chips.

$\dot{M}s$ (kg/s) of solid particles per unit time and the mass flux $\dot{M}g$ (kg/s) of the gas (air) per unit time through the cross sectional area of a pipe as

$$M = \frac{\dot{M}s}{\dot{M}g} \qquad (7)$$

Then, supposing that the mass flux of gas is equal to $\dot{M}g = Qo\,\rho$, Equation 7 can be written

$$M = \frac{\dot{M}s}{\rho \cdot Qo} \qquad (8)$$

or the flow rate of gas Qo (m³/s) can be written

$$Qo = \frac{\dot{M}s}{\rho\,M} \qquad (9)$$

IV. TRANSPORATION MECHANISM OF SOLID PARTICLES THROUGH A PIPE LINE

Here we must estimate a minimum flow velocity Vm of gas (air) which must exactly and steadily transport solid particles by gas through the pipe line. Roughly speaking, the required power for the pneumatic transportation is proportional to the third power of the gas velocity Vo. Therefore, it is undesirable to estimate the high value of the gas transportation velocity.

There are many papers concerning the transporation flow mechanism of the solid particles by gas through the horizontal pipe line, but until now it has been very difficult to be clear enough. Therefore, the author wants to explain a fundamental mechanism of solid particle transportation by gas through a pipe line. For a solid particle which is deposited on the surface of a horizontal pipe line, as shown in Figure 3, the following forces are

G(N) is the gravitational force of a solid particle
L(N) is the lift force
F(N) is the pressure force
Fs(N) is the frictional force on a solid surface

Table 2
FORMS OF SPHERICAL AND IRREGULAR PARTICLES CORRESPONDING TO THE SAME TERMINAL VELOCITY IN THE AIR CLASSIFIER (PARTICLE REYNOLDS NUMBER $10^{-3} <$ REX < 10)

Particle form	Side view	Overlook view	Height Lh/Kp	Diameter Ld/Xp	Length Ll/Xp	Sieve width Ls/Xp
Sphere			1	1	1	1
Cube			1	1.3	1.7	1.2
Rectangle			1	1.9	2	1.2
Fragment			1	2	3	1.5
Flat			0.8	3—8	4—10	2—5
Needle			1	—	3—8	1.5
Fiber			1	—	5—50	—

Note: Lh: height of the flat particle, Ld: mean diameter for overlook, Ll: maximum length, Ls: free width of sieve meshes, Xp: diameter of a spherical particle which is equivalent of the same terminal velocity.

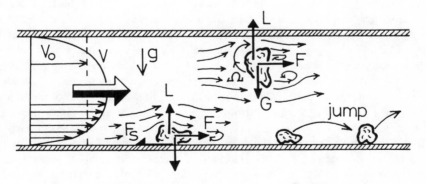

FIGURE 3.   Transportation mechanism of the solid particles through a pipe line.

Then, the most important factor is the generation of lift force L which can keep the solid particles in a floating movement after flying from the pipe's bottom. The generation of lift force L may be correlated to the influence of an unsymmetrical gas flow around the solid particles. Thereupon when the gas flows around the surface of the solid particles, as shown in Figure 4, the eddy currents are formed by the separation of the gas flow. Therefore, the lift force L, rotation Ω, and driving force F for a solid particle are invested by the gas flow pressure force.

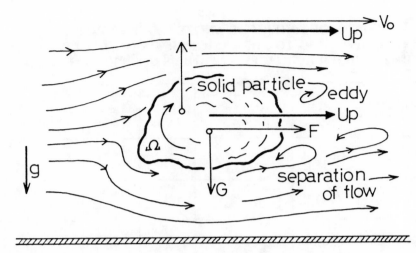

FIGURE 4.   Illustration of the aerodynamic forces acting on a solid particle in a gas flow.

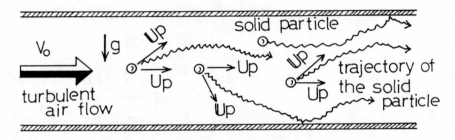

FIGURE 5.   Trajectories of solid particles in the turbulent air flow through a pipe line.

On the other hand, in order to raise solid particles from the bottom of a pipe, a low velocity of gas is not enough. This is because the pressure force F on a solid particle by the gas flow near the bottom of the pipe cannot overcome the friction force Fs of the solid particle. However, by increasing the mean gas velocity Vo, the solid particles begin to move, slide, and roll along the bottom surface of the pipe. The mean gas (air) velocity at which the solid particles begin to move along the bottom surface of the pipe is generally called as the sliding velocity, Vom. Furthermore, with increasing the mean gas velocity Vo, the lift force L increases and finally overcomes the gravitational force G of the solid particle. Therefore, the solid particle can fly from the bottom surface of the pipe. In this case, the solid particles move along the pipe line axis; the motion of the solid particles is unstable and they simultaneously fall to the bottom and rise again from the bottom of the pipe. Consequently the velocity Up of the solid particles is shown in comparison with the velocity Vo of gas. However, with increasing the gas velocity Vo, the lift force L is increased and the solid particles can be kept stable and floating in the turbulent gas (air) flow, and then move with small amplitude vibration in the horizontal pipe, as shown in Figure 5. The axial velocity Up of a solid particle along the pipe line axis begins to increase near the entrance region and begins to become a constant velocity. Figure 6 shows the experimental results of the axial velocity Up for one solid particle which is related to the mean gas (air) velocity Vo. From these experimental results, the velocity Up of the transportation for one solid particle increases with decreasing the density ρ_s of the solid particle and relates to the kinds of materials and forms of the solid particles.

In the case of the multi-particle flow system in the gas (air) flow, the velocity Up of the

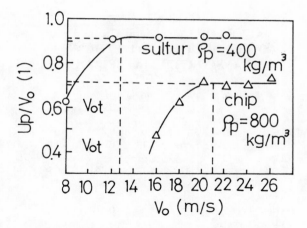

FIGURE 6. Relationship between Up and Vo for one solid particle in gas flow gas flow by Sviatkov (1966).

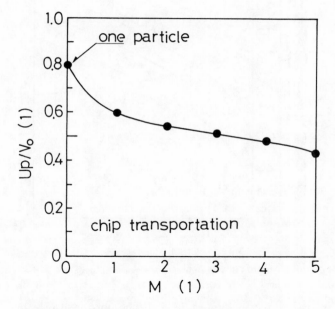

FIGURE 7. Relationship between Up and Vo for the horizontal pneumatic transportation of industrial chips.

multi-particle flow begins to decrease with the increasing concentration of multi-particles in the gas flow. The multi-particles then begin to settle on the bottom of the pipe. Increasing the mean gas (air) velocity Vo, the solid particles again begin to fly from the bottom and move with a stable motion. Therefore, in order to obtain the stable velocity Up for the multi-particle system, as for one solid particle in the horizontal pipe line, it is necessary to increase the mean gas velocity Vo. Consequently, the value of the ratio Up/Vo must be decreased.

In the case of general industrial chip transportation, the values of Up/Vo for a horizontal pipe line decrease with an increase in the mixture ratio M, as shown in Figure 7. The same experimental results were obtained by the investigation of Korobov (1974).[5] For this reason, in the actual pneumatic transportation, the transportation velocity Vot can be calculated in

Table 3
VALUES OF C_1, m_1, and C_2, m_2 IN
CONNECTION WITH THE
TRANSPORTED MATERIALS

Material	C_1	m_1	C_2	m_2
A piece of wood				
Fine particle	5.12	1.03	0.488	1.04
Rough particle	5.46	1.12	0.433	1.11
Chip				
Fine particle	5.47	1.18	0.364	1.05
Rough particle	5.46	1.25	0.380	1.11
Technical classification chip	6.15	1.30	0.410	1.25

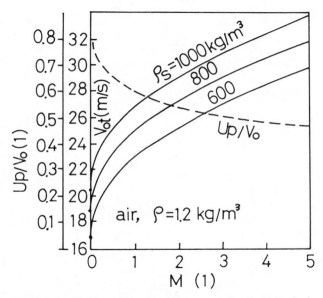

FIGURE 8. Up/Vo and Vot for chip transportation through a horizontal pipe line.

connection with increasing the mixture ratio M. In the horizontal pipe line, the result of Sviatkov (1972) is as follows

$$Vot = (C_1 M^{0.46} + m_1 \rho_s^{0.40}) \sqrt{\frac{1.2}{\rho}} \qquad (10)$$

where M is the mixture ratio (1), ρ_s is density of the chip (kg/m³), ρ is gas density (kg/m³), also C_1 and m_1 are determined by the kinds of transported materials, as shown in Table 3. For the case of the stable pneumatic transportation of the multi-particle system, the value Up/Vo can be written

$$\frac{Up}{Vo} = \frac{1}{(C_2 M^{0.46} + m_2) \cdot \sqrt{1.2/\rho}} \qquad (11)$$

where the coefficients C_2 and m_2 are determined by the kinds of the materials, as shown in Table 3. For chips of industrial classification, the values of Vot and Up/Vo were obtained, as shown in Figure 8. In this figure, when the density of gas is higher than the standard value ($\rho_o = 1.2 \, kg/m^3$), then the influence of the transportation velocity Vot can be estimated by the transportation velocity Vot, o, which is required for the transportation of solid particles through the air flow on the density ρ_o of the standard condition. A relationship between Vot and Vot, o can be written from the equilibrium state of the dynamic pressure as

$$p_{d,0} = \rho_o \cdot \frac{V_{ot,o}^2}{2} = p_d = \rho \cdot \frac{V_{ot}^2}{2} \tag{12}$$

From this equation, we can obtain the following:

$$V_{ot} = V_{ot,o} \sqrt{\frac{\rho_o}{\rho}} = V_{ot,o} \sqrt{\frac{1.2}{\rho}} \tag{13}$$

Therefore, we can transform Equation 10 with Equation 13

$$\frac{V_{ot}}{\sqrt{1.2/\rho}} = V_{ot,o} = C_1 \, M^{0.46} + m_1 \, \rho_s^{0.4} \tag{14}$$

On the contrary, for the upward transportation of solid particles through a vertical pipe line, the gas velocity Vo must keep higher values than the terminal velocity Wsg of the solid particles. Here, if we assume that the transportation velocity Vo and the concentration of the feed particles are equal values for the horizontal and vertical pipe lines, so the moving velocity of the solid particles is nearly equal in value to each other. However, in the case of the vertical pipe line, this velocity shows somewhat higher values (about 10%) in comparison with the horizontal pipe line. The reason why such a phenomenon occurs is that the concentration distribution of solid particles in the cross-sectional area is distributed nearly homogeneously and airflow always acts on all of the cross-sectional area. The value of Up/Vo for the vertical pipe line is nearly equal to the value of Up/Vo for the horizontal pipe line.

V. PRESSURE DROP FOR PNEUMATIC TRANSPORTATION

A. Horizontal Pipe Line

The motion of the solid particles through the pipe line is accompanied by collisions of particles with one another and with the surface of the pipe. From those results, since the solid particles lose their kinetic energy, the additive energy loss (pressure drop) of the gas (air) flow takes place for a supplement of the lost energy. This pressure drops more as the difference Vo − Up increases. Denoting that the pressure drop of the gas flow through the horizontal pipe line is Δp_o and that the pressure drop of the gas-solid particle flow for the same gas velocity Vo is Δp_p, this equation can be described generally as

$$\Delta p_p = \Delta p_o \, (1 + K \, M) = \lambda \cdot \frac{L}{D} \cdot (1 + K \, M) \cdot \frac{\rho \cdot V_o^2}{2} \tag{15}$$

where K is the composite experimental value. From Equation 15, the pressure drop Δp_p is increased with increasing the coefficient K and the mixture ratio. Now the value of the coefficient K is related to the gas velocity Vo, flow pattern of gas-solid motion, the diameter of pipe D, the sorts of the particles, density ρ_s of the particle, and the size and form of the particles. Figure 9 shows the relationship between the coefficient K and gas velocity Vo for the pneumatic transportation of industrial chips in the horizontal pipe line of diameter D = 260 mm.

From this experimental result, since the motion of solid particles is an unstable condition for the low velocity Vo of gas, then the value of Up/Vo is small and, therefore, the pressure

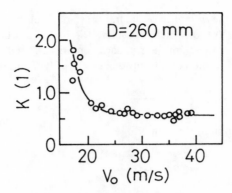

FIGURE 9.   Relationship between K and Vo
for the industrial chip transportation of the horizontal pipe line of D = 260 mm.

drop and the coefficient K are large. Increasing the velocity Vo of gas, the motion of the
solid particles gradually becomes stable, then the values of the pressure drop and of the
coefficient K are decreased. When the motion of solid particles reaches the stable state, the
value of the coefficient K becomes a constant value.

Then, in the case of the pneumatic transportation in a long length straight pipe of the
diameter D = 300 mm, Sviatkov (1966) gave the following values of K as

Piece of wood .... K = 0.82
Chip .... K = 0.78
Industrial chip .... K = 0.70
Piece of wood (silver
    fir) including the relative hu-
    midity until 65% .... K = 0.75

These values of K were obtained under the condition of stable motion for timber at density
ρ_s = 500 to 800 kg/m³. With increasing the pipe diameter, the difference of solid particle
concentration between the upper and bottom part of a pipe at an arbitrary cross-section is
increased. Therefore, it leads to the increment of the value of K. A relationship between K
and D was given by Dziadzio (1961)[6] in the case of the low feed concentration as

$$K_2 = K_1 \left(\frac{D_2}{D_1} \right)^n \qquad (16)$$

where K_1 corresponds to the diameter D_1 of a pipe and K_2 is an unknown quantity which
corresponds to the diameter D_2 of a pipe. An index n will vary between 0.5 and 1.0, but
the value of n = 0.5 to 0.7 is desired.

B. Vertical Pipe Line

In the case of the upward vertical pneumatic transportation against a gravity force, the
pressure drop Δp_v for the vertical pipe line can be roughly estimated as follows:

$$\Delta p_v = \Delta p_o \cdot (1 + K \cdot M) + \Delta p_{ov} \qquad (17)$$

where $\Delta p_o \cdot (1 + K \cdot M)$ is the pressure drop corresponding to the equivalent horizontal length
h of the pipe. The pressure drop Δp_{ov} corresponding to the upward vertical pneumatic

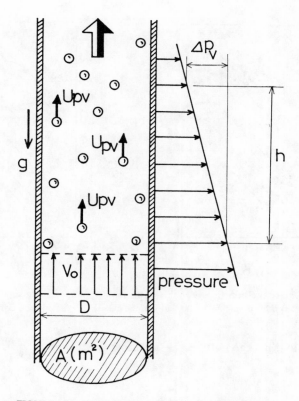

FIGURE 10.   Illustration of the vertical upward transportation
of solid particles in gas flow.

transportation for vertical length $\underline{h}$ can be derived by the following idea, as shown in Figure 10. So the mechanical equilibrium equation for the vertical length $\underline{h}$ can be written

$$\rho \cdot g \cdot h \cdot A + Mpu \cdot g \cdot h \ = \ \Delta p_{ov} \cdot A \qquad (18)$$

where Mpu is the mass of the solid particle per unit length as a unit kg/m and A is the cross-sectional area (m²) of the pipe. Here, a relationship between the mass flux $\dot{M}p$(kg/s) of the solid particle and Mpu (kg/m) can be written

$$Mpu \ = \ \frac{\dot{M}p}{Upv} \qquad (19)$$

where Upv is the mean vertical velocity of the solid particles. Substituting Equation 19 into Equation 18, the following equation can be obtained

$$\rho \cdot g \cdot h \cdot A + \frac{\dot{M}p}{Upv} \cdot g \cdot h \ = \ \Delta p_{ov} \cdot A \qquad (20)$$

Therefore, the equation of Δp_{ov} can be written

$$\Delta p_{ov} \ = \ \rho \cdot g \cdot h + \frac{\dot{M}p \cdot g \cdot h}{Upv \cdot A} \qquad (21)$$

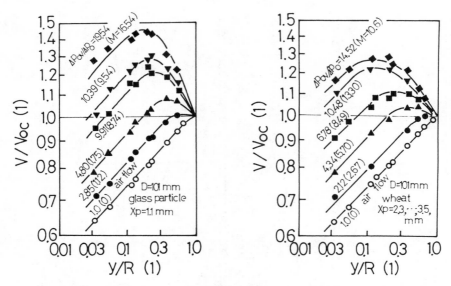

FIGURE 11.   Velocity distributions of air flow on the vertical upward transportation of solid particles in the pipe (Voc is the gas velocity at the center of the pipe, P_{ov} is the pressure drop on the mixture ratio M).

Here, the mass flux Ma (kg/s) of the gas (air) flow is equal to

$$\dot{M}a = \rho \cdot A \cdot Vo \qquad (22)$$

and the mixture ratio M is equal to

$$M = \frac{\dot{M}p}{\dot{M}a} \qquad (23)$$

then Equation 21 can be transformed to

$$\Delta p_{ov} = \rho \cdot g \cdot h + \frac{\rho \cdot M \cdot g \cdot h \cdot Vo}{Upv} \qquad (24)$$

Now in order to facilitate the above equation, changing the value g = 9.81 m/s² to g ≑ 10 m/s², finally we can obtain the following equation

$$\Delta p_{ov} \doteqdot 10 \cdot \rho \cdot h \cdot \left\{ 1 + \frac{M}{(Upv/Vo)} \right\} \qquad (25)$$

where the value of Upv/Vo can be roughly estimated from Equation 11.

On the other hand, comparing the inclined pipe line of the angle θ against the horizontal pipe line, the pressure drop for the length l (m) for the upward pneumatic transportation can be applied in the following equation

$$h = l \sin \theta$$

Figure 11 shows the velocity distributions of air in the vertical upward transportation of the solid particles (glass particles and wheat) in the pipe of diameter D = 101 mm by Vollheim.[7,8] In this figure Voc is the velocity of gas (air) at the pipe's center, V is the velocity of gas,

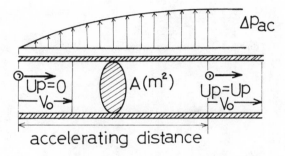

FIGURE 12. Illustration for the acceleration of solid particles through a horizontal pipe line.

y is the distance from the wall of the pipe, R is the radius (= D/2), Δp_O is the pressure drop on the pure air flow, and ΔP_{OV} is the pressure drop on the solid particles moving upward in the pipe. From these figures, the velocity distributions of gas (air) on the vertical upward flow in the pipe (D = 101 mm) deviated from a logarithmic velocity distribution

$$\frac{V}{V_*} = 2.5 \ln \frac{V_* y}{v} + 5.5$$

with increasing the mixture ratio M, where V_* is the frictional velocity.

VI. LOCAL ADDITIVE PRESSURE DROP

Local additive pressure drop (for example, flow through the bend) Δp_{ol} for the pneumatic transportation in straight pipe line can be calculated as

$$\Delta p_{pl} = \Delta p_{ol} (1 + K \cdot M) = (1 + K \cdot M) \, \zeta \cdot \rho \cdot \frac{V_o^2}{2} \qquad (26)$$

where Δp_{pl} is the local additive pressure drop for motion of solid particles and Δp_{ol} is the local additive pressure drop of the air flow which can be described as

$$\Delta p_{ol} = \zeta \cdot \rho \cdot \frac{V_o^2}{2} \qquad (27)$$

and also $(1 + K \cdot M)$ is the numerical factor of the pneumatic transportation and ζ is the local friction coefficient of the pure air flow. For example, pneumatic transportation in the pipe line with bends was investigated by the computer simulation method by Tsuji and Morikawa.[9]

VII. ACCELERATION OF SOLID PARTICLES

In the transportation of solid particles through the pipe line, the solid particles must be accelerated to the constant velocity under the turbulent air flow.[10] Thererfore, a part of the fluid energy must be compensated as a pressure drop. Then, the distance completing the acceleration of the solid particles is called an accelerating distance, as shown in Figure 12. Assuming that the solid particle increment of the momentum is equal to the impulse force of the fluid flow on the accelerating distance, we can obtain the following equation

$$\text{Fac} \cdot t = Mp \cdot (Up - 0) = Mp \cdot Up \qquad (28)$$

where $\text{Fac} = \Delta p_{ac} \cdot A$ is the fluid force acting on the solid particles at the accelerating distance, t is an accelerating time, Mp is the particle mass, and Up is the attainable velocity

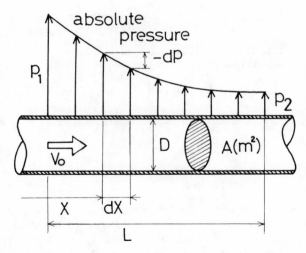

FIGURE 13.   Illustration of the pressure drop of gas flow along
the pipe line.

of the solid particle after the end of the acceleration. From Equation 28 we can obtain the
equation of the pressure drop ΔP_{ac} as

$$\Delta p_{ac} = \frac{Mp \cdot Up}{A \cdot t} \tag{29}$$

Multiplying a factor $2 \cdot \rho \cdot Vo^2$ to the numerator and the denominator of the right hand side of
Equation 29, we can then obtain the following equation

$$\Delta p_{ac} = \frac{\dot{M}p \cdot 2 \cdot \rho \cdot \overset{2}{Vo} \cdot Up}{\dot{M}a \cdot 2 \cdot Vo} \tag{30}$$

where $Mp/t = \dot{M}p$ is the mass flow of the solid particles and $\rho \cdot A \cdot Vo = \dot{M}a$ is the mass
flux of the gas (air) flow. Now substituting the mixture ratio $M = \dot{M}p/\dot{M}a$ into Equation
30, Equation 30 can be transformed to

$$\Delta p_{ac} = 2 \cdot \rho \cdot M \cdot \left(\frac{Up}{Vo}\right) \cdot \frac{\overset{2}{Vo}}{2} \tag{31}$$

where the value of Up/Vo can be calculated from Equation 11.

VIII. PRESSURE DROP IN THE REGION OF GAS EXPANSION ALONG THE PIPE LINE

In the case of a pressure drop of more than 20,000 Pa by pneumatic transportation, the
calculation of pressure drop must consider the expansion of gas (air) along the pipe line.
Therefore in the flow process of gas (air) in a pipe line, the mass flux of the air flow is at
constant value, but the volumetric flow rate is increased. Consequently, the velocity of gas
(air) through the pipe line must be increased, and hence the pressure drop occurs.

As shown in Figure 13, assuming that the pressure drop (-dp) along the small distance
dx of a pipe diameter D can be written

$$-dp = \lambda \cdot \frac{1}{D} \cdot d\left(\rho \cdot x \cdot \frac{\overset{2}{Vo}}{2}\right) \tag{32}$$

we can obtain the equation of the pressure drop Δp. Here, we may regard the coefficient of the pipe friction λ as a constant value in the fully turbulent air flow. Now denoting that R is gas constant (kJ/kg·K), T is the absolute temperature (K) of gas, and p is the absolute pressure (Pa) of gas, the mean gas velocity can be written

$$V_O = \frac{\dot{Mg} \cdot R \cdot T}{A \, p} \qquad (33)$$

Substituting Equation 33 into Equation 32, we can obtain the fundamental equation

$$-dp = \frac{\lambda}{D} \cdot dx \left(\frac{Mg}{A}\right)^2 \cdot \frac{R \cdot T}{2 \cdot p} \qquad (34)$$

Therefore, Equation 34 can be transformed as

$$-\int_{p_1}^{p_2} p \, dp = \int_0^L \frac{\lambda}{D} \cdot \left(\frac{\dot{Ma}}{A}\right)^2 \cdot \frac{R \cdot T}{2} \cdot dx \qquad (35)$$

then integrating Equation 35, we can obtain the following equation

$$\frac{p_1^2 - p_2^2}{2} = \frac{\lambda}{D} \cdot \left(\frac{\dot{Ma}}{A}\right)^2 \cdot \frac{R \cdot T \cdot L}{2} \qquad (36)$$

Here, using the value of R (air) = 287 J/kg·K $\fallingdotseq$ 300 J/kg·K, Equation 36 is transformed to

$$p_1 \fallingdotseq \sqrt{p_2^2 + 300 \cdot \frac{\lambda}{D} \cdot \left(\frac{\dot{Ma}}{A}\right)^2 \cdot T \cdot L} \qquad (37)$$

Now, when the geometrical length of the pipe is Lg and the equivalent length by the local additive energy loss is ΣLad, the length L of the right hand side of Equation 37 can be written

$$Le = Lg + \Sigma Lad \qquad (38)$$

Consequently, the pressure drop which contains the local additive pressure drop can be written

$$\frac{\lambda}{D} \cdot Le \cdot \rho \cdot \frac{V_O^2}{2} = \zeta \cdot \rho \cdot \frac{V_{oe}^2}{2} \qquad (39)$$

where ζ is the local friction coefficient and Voe, Vo are the air velocities at the cross-sectional area of the local resistant portion and of the transportation pipe. Therefore, from Equation 39 we can obtain

$$Le = \frac{\zeta}{(\lambda/D)} \left(\frac{V_{oe}}{V_o}\right)^2 \qquad (40)$$

If we assume that Voe is nearly equal to Vo, then Equation 40 can be written

$$Le = \frac{\zeta}{(\lambda/D)} \qquad (41)$$

Therefore, in the region of the local additive resistant domain, Equation 37 can be rewritten as

$$p_1 = \sqrt{p_2^2 + 300 \cdot \frac{\lambda}{D} \cdot \left(\frac{\dot{Ma}}{A}\right)^2 \cdot T \cdot Le} \qquad (42)$$

Then it may be possible to write the pressure drop for the pneumatic transportation of the solid particles from Equation 42 as

$$p_1 = \sqrt{ p_2^2 + 300 \cdot \frac{\lambda}{D} \cdot \left(\frac{\dot{M}a}{A} \right)^2 \cdot T \cdot Le \cdot (1 + K \cdot M)}$$ (43)

Also the total pressure drop Δp which contains the acceleration and the lift force for the above considered region can be written

$$\Delta p = \sqrt{ p_2^2 + 300 \cdot \frac{\lambda}{D} \cdot \frac{\dot{M}a}{A}^2 \cdot T \cdot Le \cdot (1 + K \cdot M)}$$

$$+ \Delta p_{ov} + \Delta p_{ac} - p_2$$ (44)

REFERENCES

1. **Taber, B. A., Kaliiteevskij, R. E., and Gromchev, E. K.,** *Vnutrizabodskoj Transport,* Lesnoa Prompychlennocte, Moskva, CTb 1978.
2. **Morikawa, Y.,** Examples of the systematic design for the low velocity-high concentration pneumatic transportation abroad, *J. Soc. Powder Tech. Jpn.,* (in Japanese), 18(11), 831, 1981.
3. **Morikawa, Y.,** *Two-Phase Flow of Fluid-Solid Particles,* (in Japanese), Nikan-Kogyo Shinbunsha, Tokyo, 1979.
4. **Tzuji, Y. and Morikawa, Y.,** Plag flow of coarse particles in a horizontal pipe, Proc. Symp. Polyphase Flow and Transport Tech., San Francisco, 1980.
5. **Korobov, V. V.,** *Pnevmatitcheskij Transport i Pogruzka Tehnologitcheskoj Shchepy,* 1974, 176.
6. **Dziadzio, A. M.,** *Pnevmatitcheskij Transport na Zernopererabatyvaiushchih Predpriiatiiah,* 1961, 328.
7. **Vollheim, R.,** Verhalten der Wandschubspannung des Fördermediums beim pneumatischen Transport und Schlußfolgerungen für den Wärmeübergang, *Maschinenbautechnik,* 12(5), 233, 1963.
8. **Vollheim, R.,** *Pneumatischer Transport,* VEB-Deutscher Verlag, Leipzig, 1971.
9. **Morikawa, Y. and Tzuji, Y.,** Computer simulation for the pneumatic transport in pipes with bends, *Pneumotransport,* 4, 1978.
10. **Fortier, A. and Chen, C. P.,** Écoulement turbulent stationnaire biphasique air-solide dans un tube cylindrique, à forte concentration massique, *J. Mec.,* 15(1), 155, 1976.

Chapter 8

INERTIAL SEPARATOR

I. INTRODUCTION

In this chapter the inertial forces of dust or solid particles in a flowing gas are applied to the several types of dust collectors.

It is very difficult to separate solid particles under the particle size $Xp \fallingdotseq 30$ μm. Those inertial separators can be applied for precollectors or for dust concentrators. As an example of this device, Figure 1 shows the baffled settling chamber which used the control baffle for re-circulating the flow after it entered this device and the pipe cleaning baffle.[1,2] Although most of the gas flows upward between the pipe cleaning baffles to the exit, coarse particles and residual gas flow through the dust slot at the apex of the passage. In the re-circulating flow region, the dust-laden gas accepts the centrifugal force on the circulating flow region and also the gravitational force. By those two types of forces, coarse particles are separated easily. After that, when the small particles enter the control baffle, the loci of the solid particles changes the direction and the particles impinge on the surface of the baffle. The collection efficiency η_x (Xp) for fly-ash of $Xp \fallingdotseq 34$ μm is about 50% as shown in Figure 2 by Ambuco.

On the other hand, many types of the baffle (louver) designs are as shown in Figure 3. The collection efficiency η_c is, in general, dependent on the degree of impingement which occurs. Baffled chamber efficiency may be a function of four variables: the number, length, spacing, and configuration of the baffles.

II. LOUVER DUST COLLECTOR WITH CYCLONE

Figure 4 (Muschelknautz[3]) shows a louver dust collector which is connected with subsidiary cyclones and a bag filter. The diameter of louver collector in this case is $D_1 = 1100$ mm. About 90% of air in the dust-laden gas flows through the conical slits and turns about 180° (π rad); at the same time the coarse dust particles are separated from gas as shown by a schematic diagram in Figure 4.

Figure 5 shows the loci of solid spherical particles of diameter $Xp = 10$ μm and of density $\rho_p = 1.3$ g/cm³ which correspond to the inertia parameter Vo·Wsg/g·S = 1.74. The stream lines include a stream line of a stagnation point for Vo = 22.5 m/s. The loci of solid particles are calculated by the assumptions of Stokes drag force and neglect the gravity force of the solid particle. A locus of the left line for the solid particle is turned about 180°, therefore this particle cannot be separated. However, a locus of the right line for the solid particle can be separated. Then a locus of a middle line for the solid particle may collide with the wall, consequently to be separated or not to be separated for the solid particle reaching the stagnation point is a probability problem.

Therefore, if the dust-laden gas of diameter $Xp = 10$ μm is distributed homogeneously in this louver collector, about 80% of these particles may be separated. Figure 6 shows the experimental results (Co = 50 to 150 g/m³) of the fractional collection efficiency η_x which is a function of inertia parameter defined as

$$\frac{Vo \cdot Wsg}{g \cdot S} = \frac{Vo \cdot (\rho_p - \rho) \cdot Xp^2}{18 \cdot \eta \cdot S} \tag{1}$$

In this figure, symbol S means the distance of a slit as shown in Figure 5, ρ_p and ρ means

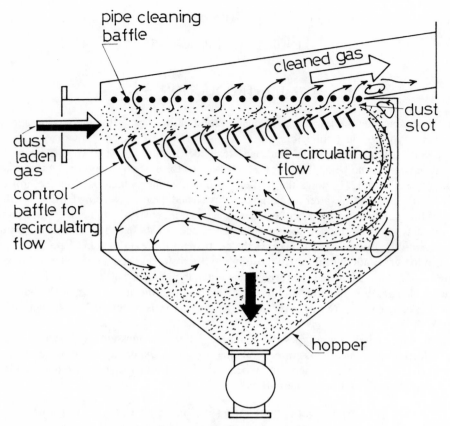

FIGURE 1.   Baffled settling chamber.

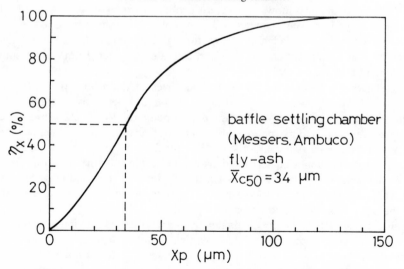

FIGURE 2.   Fractional collection efficiency of baffle settling chamber for fly ash.

the density of particle and gas, respectively. This equation may be derived from the following mechanical consideration

$$\frac{\text{centrifugal force}}{\text{Stokes drag force}} = \frac{(\rho_p - \rho) \cdot \dfrac{\pi \cdot X_p^3}{6} \cdot \dfrac{V_o^2}{S}}{3 \cdot \pi \cdot \eta \cdot X_p \cdot V_o} = \frac{V_o \cdot W_{sg}}{g \cdot S}$$

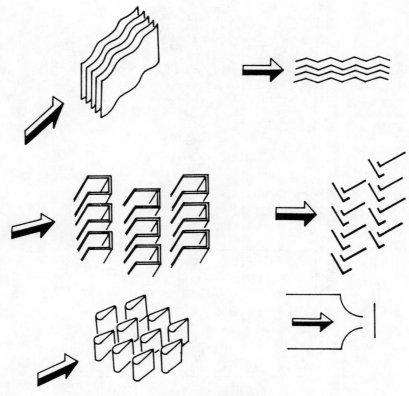

FIGURE 3.   Baffle designs.

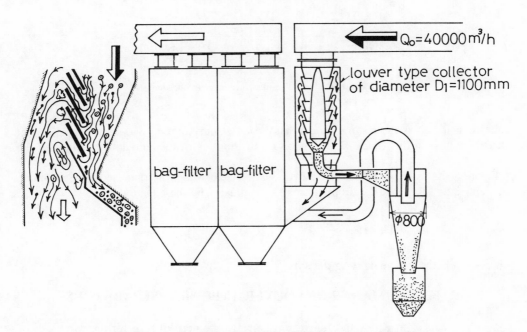

FIGURE 4.   Louver dust collector with subsidiary cyclone and bag filter.

where Wsg is a terminal velocity of the particle in gas.

Figure 7 shows the fractional collection efficiency $\eta_x(Xp)$ for louver and cyclone dust

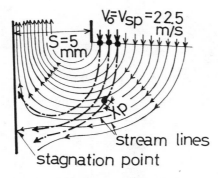

FIGURE 5. Stream lines and particle loci in louver type of collector for Xp = 10 μm, ρ_p = 1.3 g/cm³, Wsg = 0.392 cm/s, VoWsg/S · g = 1.74.

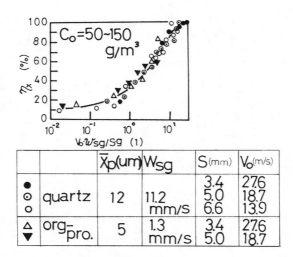

		$\overline{X}p$(um)	Wsg	S(mm)	Vo(m/s)
● ⊙ ○	quartz	12	11.2 mm/s	3.4 5.0 6.6	27.6 18.7 13.9
△ ▼	org-pro.	5	1.3 mm/s	3.4 5.0	27.6 18.7

FIGURE 6. Experimental results of fractional collection efficiency of Figure 5.

collectors. It is very interesting to note that those two curves of η_x(Xp) show very different characteristics. In this figure, R(Xp) shows a distribution of the cumulative residue of the particles.

Figure 8 shows one example of a louver dust collector which is connected with a cyclone dust collector. The pressure drop Δp_c may be estimated by the equation

$$\Delta p_c = (3 - 4) \cdot \frac{\rho \cdot V_{sp}^2}{2} \tag{2}$$

where Vsp is the gas velocity in the slits.

III. PERFORMANCE OF LOUVER-TYPE MIST SEPARATORS

Ushiki, Takahashi, Kato, and Iinoya[4] investigated the performance for the two kinds of louver type mist separators. One type has V-shaped blades and another type has flat blades, as shown in Figure 9. Those blades are set on the slant to the mist-laden gas flow. The distribution of the particle size Xp of the salt-water droplets distributes between Xp = 10 to 40 μm and the mode diameter is about Xp = 20 μm.

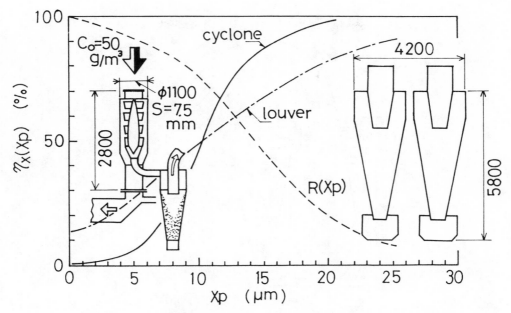

FIGURE 7.   Fractional collection efficiency for louver and cyclone dust collectors.

Figure 10 shows the relationship between the fractional collection efficiency $\eta_x(Xp)$ and the particle diameter Xp for the V-shaped blades without blow down. The parameter in this figure is the mean air velocity Vo. Increasing air velocity Vo, the fractional collection efficiency η_x is also increased. The cut-size Xc corresponding to $\eta_x(Xp) = 0.5$ for Vo = 120 cm/s is about 13.5 μm.

Figure 11 shows a relationship between the fractional collection efficiency $\eta_x(Xp)$ and the inertia parameter Ψ based upon the blade distance B defined as

$$\Psi = \frac{\rho_p \cdot Vo \cdot Xp^2}{18 \cdot \eta \cdot B} \tag{3}$$

for the flat blades with 10% blow down flow. In order to compare the separation performance for the droplets and the solid particles, the experimental results of $\eta_x(Xp)$ for the salt-water droplets and for the fly-ash are shown. From this figure the motion of solid particles and droplets after the first collision with the blade shows very different behavior. For instance, in the case of the solid particles, some of them may be collected after repeating several collisions with the blade, but another part of the particles may pass through the blades due to the weak adhesion force.

In the case of the droplets after a collision with the blade, a thin layer of liquid flows down to the surface of the blades and begins to fall to the dust bunker until there are several millimeters of liquid droplets at the edge of the blade. Therefore, the collection efficiency of the droplets is better than that of the solid particles.

Usiki et al. reported that in the case of blow-down flow, the collection efficiency of a V-shaped blade is higher than that of a flat blade. The former is about twice the latter when the latter is 40%, even though the pressure drop of the V-shaped blade is only three fourths that of the flat blade. Further, there is little difference in collection efficiency between V-shaped and flat blades under the condition of a 10% blow-down flow, in the case of the solid particles.

Figure 12 shows the relationships between the pressure drop Δp_c and the inlet air velocity Vo. The pressure drop Δp_c can be written

$$\Delta p_c = k \cdot Vo^m \tag{4}$$

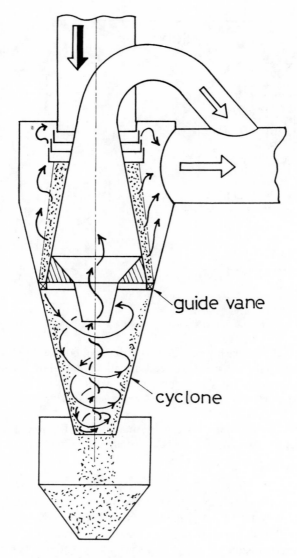

FIGURE 8.   One example of a louver dust collector.

where an exponent m is in the limits of m = 1.7 to 1.9. The pressure drop of flat blades is larger than that of the V-shaped blades. This tendency coincides with the experimental results of Gee and Cole (1969—1970).

IV. INERTIA SEPARATOR

This type of separator does not separate enough in comparison with the centrifugal separator, but can be applied in the separation of corn.[5] This type of separator is also used for the separation of crushed fireproof material and shows good separation efficiency in comparison with the centrifugal separator.

An outline of this separator is shown in Figure 13. Air is introduced from pipe 1 to pipe 2 (in a zig-zag fashion) which is connected with pipe 3. Pipe 3 is connected with pipe 4, which is fed the materials. Then pipe 3, by way of pipe 5, is connected with pipe 6 at an

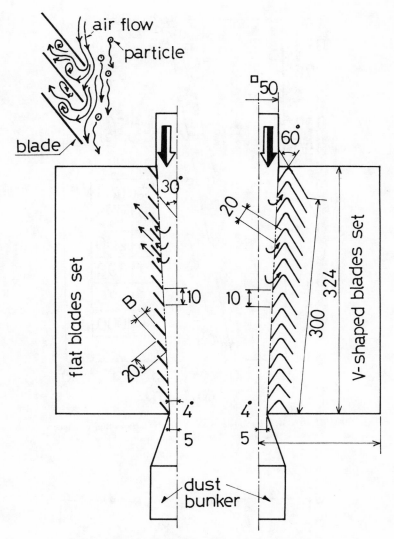

FIGURE 9. Two kinds of louver type mist separators.

elbow. On the upper portion of pipe 6, exhaust pipe 7 is equipped, and at the lower portion of pipe 6, pipe 8, air is introduced. Pipes 3 and 6 are connected to bunkers 9 and 10.

The separation mechanism is as follows: air introduced from pipes 1 and 8 flows up along pipes 3 and 6. The material introduced from pipe 4 is fed to pipe 3 and then is classified as coarse and small material by the upward air flow. Coarse particulate material moving against the upward air flow falls to pipe 2 of the zig-zag element. Then, in this element, an eddy motion of fluid promotes the separation of the coarse particulate material from the small material which is transported to pipe 3 by the upward air flow. Further, the coarse material is collected to bunker 9. On the other hand, the small particulate material is transported from pipe 3 to pipe 6 (zig-zag way) by way of pipe 5. In this pipe element 6, the separation to the small and fine material occurs. By the upward air flow entering from pipe 8, small particle materials are separated from the fine particulate materials. Then small particulate materials are carried to bunker 10. Therefore, fine particle materials are transported to pipe 7 and escape from this separator. For the fireproof materials the separation efficiency η_c can be obtained about 90% over for the material load 200 kN/h and for the 1 m² sectional area of the vertical pipe.

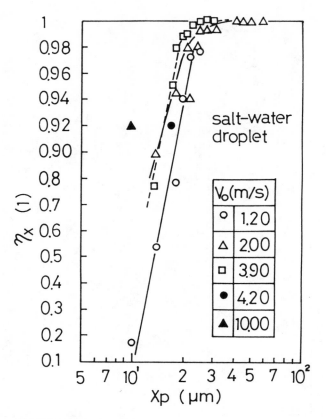

FIGURE 10.   Relationship between the fractional collection efficiency and the particle size for the V-shaped blades without blow down.

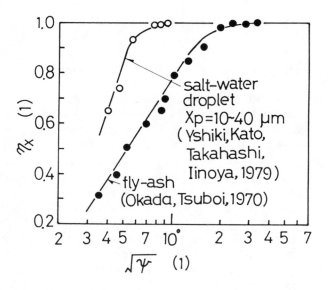

FIGURE 11.   Relationship between the fractional collection efficiency and the inertia parameter.

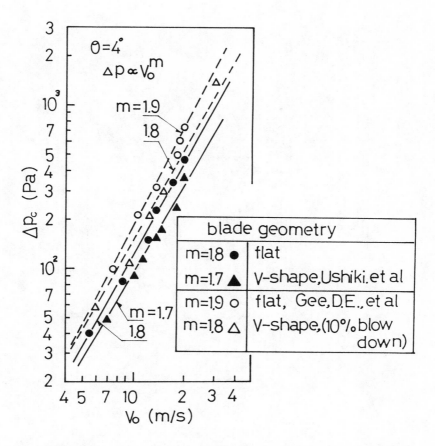

FIGURE 12. Relationship between the pressure drop and the inlet air velocity.

V. APPLICATION OF FLUIDICS

Okuda and Yasukuni[6] made a new classifier apparatus which applies the principle of Coanda of fluidics for sharply separating micron or sub-micron fine solid particles. The new apparatus is composed of a sample supplying device with a main jet flow of fluidics and a collecting device to separate samples, as shown in Figure 14.

They applied the fluid velocity distribution along the surface of half-cylinder proposed by Kimura and Mitsuoka (1968) and defined in the mixing region as

$$V = V_{max}\left\{1 - 3 \, |Y^*|^2 + 2 \, |Y^*|^3\right\} \tag{5}$$

where $Y^* = y/y_{max}$ and $V_{max} = V_o\sqrt{X_o/X}$.

Here V_{max} is the velocity of a centerline of the air stream, V_o is the velocity of nozzle outlet, X is the distance from the outlet of nozzle, and X_o is the end of transition region. Figure 15 shows one example of the velocity distribution from the nozzle outlet. The centerline of the jet can be expressed by the following equation

$$r = \left(r_o + \frac{Ws}{2}\right) + 0.081 \cdot \theta^2 \tag{6}$$

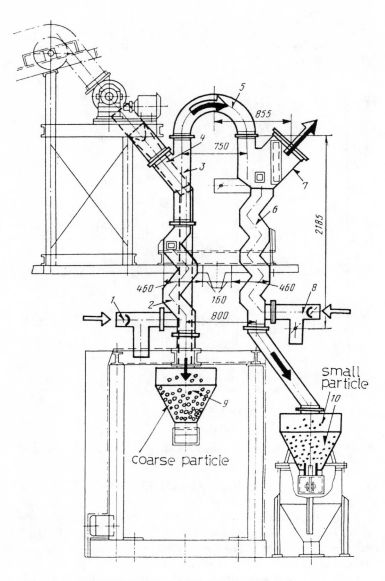

FIGURE 13.   One system of the inertia separator.

where r_o is the radius of the cylinder, r is a distance from a pole, θ is swirl angle, and Ws is the width of nozzle.

Figure 16 shows the trajectories of the particle size Xp in the range 0.5 to 5 μm for the nozzle ejecting velocity 50 m/s. In this case, all particles less than 5 μm follow up to the swirl air stream around the cylinder. From those experimental results, they found that the diverging position by knife edge is very important.

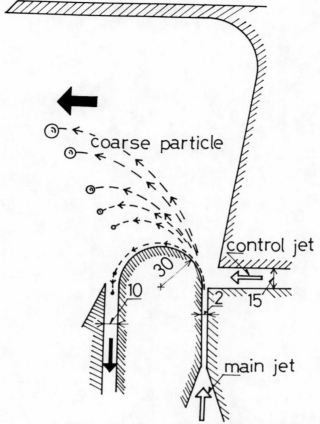

FIGURE 14.   New apparatus of the classifier.

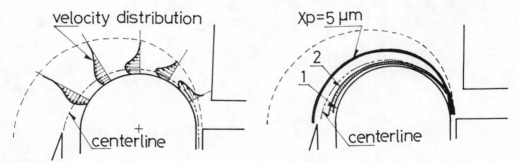

FIGURE 15.   Velocity distribution from the nozzle outlet.

FIGURE 16.   Loci of the solid particles for the nozzle ejecting velocity 50 m/s.

REFERENCES

1. **Bethea, R. M.,** *Air Pollution Control Technology,* Van Nostrand, New York, 1978.
2. **Dorman, R. G.,** *Dust Control and Air Cleaning,* Pergamon Press, Elmsford, N.Y., 1974.
3. **Muschelknautz, E.,** Untersuchungen an Fliekraftabscheidern, *Chemie Ing. Tech.,* 39(5/6), 306, 1967.
4. **Ushiki, K., Kato, H., Takahashi, S., and Iinoya, K.,** Performance of Louver Type Mist Separators, *Kagaku-Kogaku Ronbun Shy,* (in Japanese) 5(4), 348, 1979.
5. **Bajsorolov, V. G.,** Mehanitcheskoe i Transportnoe Oborudovanie Zabodoe Ogneupornoj Promyshlennosti, *Metallurgĩa,* 1981.
6. **Okuda, S. and Yasukuni, J.,** Application of Fluidics Principle to Fine Particle Classification, Int. Symp. Powder Technol., Kyoto, Japan, 1981.

APPENDIX OF MAIN SYMBOLS

CHAPTER ONE

D1 (m)	Cyclone diameter
Dp (m)	Diameter of a cylinder or a sphere
Ep (V/cm)	Intensity of the electric field
p (Pa)	Pressure of gas or air
Qo (m^3/s)	Flow rate of gas
R(Xp) (1)	Cumulative residue distribution of the particle size
r (m)	Radius
T (K)	Temperature
Up (m/s)	Migration velocity
Vθ (m/s)	Tangential velocity of gas flow
Vo (m/s)	Mean representative velocity of gas flow
Wsg (m/s)	Terminal velocity of the solid particle
Xc (μm)	Cut-size
Xp (μm)	Particle size
η (Pa·s)	Viscosity of gas
$η_c$ (1)	Total collections efficiency
$η_x$ (1)	Fractional collection efficiency
$η_t$ (1)	Target efficiency
υ (m^2/s)	Kinetic viscosity of gas
$ρ_p$ (kg/m^3)	Particle density
Φ (W)	Power
Φ (1)	Centrifugal effect

CHAPTER TWO

A (m^2)	Cross-sectional area
C_D (1)	Drag coefficient
D (m)	Diameter of pipe
f (Hz)	Vortex shedding frequency
Foτ (1)	Fourier number
Ga (1)	Galilei number
L (m)	Pipe length
m (m)	Hydraulic mean depth
Δp (Pa)	Pressure drop
Qo (m^3/s)	Flow rate of fluid
r (m)	Radius
Re (1)	Reynolds number
Re*L* (1)	Reynolds number for the flat plate
Rep (1)	Reynolds number around a particle
S̲ (1)	Strouhal number
V (m/s)	Mean velocity of gas
Vo (m/s)	Velocity at the center of pipe
X (m)	Distance
δ (m)	Thickness of the laminar sublayer
ε (m)	Surface roughness of pipe
η (Pa·s)	Viscosity of gas

Λ (m)	Wavelength
λ (1)	Pipe friction factor
υ (m²/s)	Kinematic viscosity
ρ (kg/m³)	Density of gas
τ (1)	Dimensionless time
τ (Pa)	Shear stress

CHAPTER THREE

C (1)	Drag coefficient
Cv (1)	Volume concentration of solid particles
D (N)	Drag force
Rex (1)	Reynolds number around a solid particle
Up (m/s)	Particle velocity
Vo (m/s)	Gas velocity
Wsg (m/s)	Terminal velocity of one solid particle
Wsgm (m/s)	Terminal velocity of the group of particles
Xp (μm)	Diameter of the solid particle
ρ (kg/m³)	Density of gas
ρ_p (kg/m³)	Density of the solid particle

CHAPTER FOUR

C_D (1)	Drag coefficient
D (m²/s)	Turbulent eddy diffusivity
Dp (m²/s)	Diffusion coefficient of solid particles
m (m)	Hydraulic mean radius
mp (kg)	Mass of a solid particle
Qo (m³/s)	Flow rate of gas
Re (1)	Reynolds number of flow
Rex (1)	Particle Reynolds number
Up (m/s)	Particle velocity
Vo (m/s)	Mean velocity of gas
$\sqrt{\overline{v^2}}$ (m/s)	Turbulent fluctuating velocity of gas
Vr (m/s)	Radial velocity of gas
Vθ (m/s)	Tangential velocity of gas
Wsg (m/s)	Terminal velocity of a solid particle
Xp (μm)	Particle diameter
Xc (μm)	Cut-size
υ (m²/s)	Kinematic viscosity
υ_T (m²/s)	Apparent kinematic eddy viscosity

CHAPTER FIVE

C_D (1)	Drag coefficient
D (N)	Drag force
D1 (m)	Diameter of a cyclone of system I
D'1 (m)	Diameter of a cyclone of system II
Fr (1)	Froude number
Qo (m³/s)	Flow rate of fluid
Re (1)	Reynolds number of fluid flow

Sa (1)	Similarity of the accelerations of the dust particle
S1 (1)	Geometrical similarity
St (1)	Similarity of lapse of time
Sst (1)	Stokes number, inertia parameter
Su (1)	Similarity of the loci of the dust particles
Sv (1)	Similarity of the fluid flow
t (s)	Time of system I
t' (s)	Time of system II
Vo (m/s)	Flow velocity of system I
Vo' (m/s)	Flow velocity of system II
Wsg (m/s)	Terminal velocity of the solid particle
Xp (μm)	Particle diameter
ρ (kg/m^3)	Density of gas
ρ_p (kg/m^3)	Density of the solid particle
τ (s)	Representative time for the terminal velocity

CHAPTER SIX

D (m^2/s)	Coefficient of diffusion
f(Xp) (%/μm)	Frequency distribution of R(Xp)
m (1)	Separation index
n (1)	Characteristics of the size distribution of R(Xp)
R(Xp) (1)	Cumulative distribution of the particle size
T (1)	Separation parameter.
Wsc (m/s)	Centrifugal sedimentation velocity
Xp (μm)	Particle size
Xc (μm)	Cut-size
X_{R50} (μm)	Particle size corresponding to R(Xp) = 0.5.
η_c (%)	Total collection efficiency
η_x (Xp) (1)	Fractional collection efficiency

CHAPTER SEVEN

C_D (1)	Drag coefficient
D (mm)	Pipe diameter
D (N)	Drag force
G (N)	Gravitational force
L (N)	Lift force
M (1)	Mixture ratio
$\dot{M}g$ (kg/s)	Mass flux of gas
$\dot{M}s$ (kg/s)	Mass flux of solid particles
p (Pa)	Pressure
Δp (Pa)	Pressure drop
ΔP_{ac} (Pa)	Pressure drop for accelerating distance
ΔP_{ov} (Pa)	Pressure drop for the vertical pipe
Qo (m^3/s)	Flow rate of gas
R (kJ/kg·K)	Gas constant
T (K)	Gas temperature
Up (m/s)	Particle velocity
Vo (m/s)	Fluid velocity
Wsg (m/s)	Terminal velocity of the solid particle

Xp (mm)	Particle diameter
ζ (1)	Local friction factor
λ (!)	Friction factor
ρ (kg/m³)	Density of gas
ρ_s (kg/m³)	Density of the solid particle

CHAPTER EIGHT

B (m)	Blade distance
Co (g/m³)	Dust concentration
Δp_c (Pa)	Pressure drop
S (m)	Width of the slits
Vo (m/s)	Representative velocity of the fluid flow
Vsp (m/s)	Gas velocity in the slits
Wsg (m/s)	Terminal velocity of the solid particle
Xp (μm)	Particle size
η (Pa·s)	Viscosity of fluid
η_c (%)	Collection efficiency
η_x (1)	Fractional collection efficiency
ρ (kg/m³)	Density of fluid
ρ_p (kg/m³)	Density of the solid particle
Ψ (1)	Inertia parameter

INDEX